My Father,
THE CAPTAIN

Praise for *My Father, the Captain*

"I WOULD SAY THAT CAPTAIN COUSTEAU was the 'father' of the environmental movement and Rachel Carson would be my choice for 'mother.' The Captain and the crew of the *Calypso* fascinated, informed, entertained, and educated us for decades. For me, he is sorely missed."

—Ted Turner, chairman of Turner Enterprises, Inc.
and author of Call Me Ted

"I'VE KNOWN JEAN-MICHEL FOR DECADES and thought that I knew the whole Cousteau saga fairly well, but *My Father, the Captain* was a genuine surprise! There are many published works that purport to be biographies of Jacques-Yves Cousteau and his incredible career, but, I guarantee you, this is *the* Cousteau story. I applaud Jean-Michel's generosity in sharing this intimate story with us."

—Dr. Phil Nuytten, president and founder
of Nuytco Research Ltd.

My Father,
THE CAPTAIN

My Life with
Jacques Cousteau

JEAN-MICHEL COUSTEAU

With Daniel Paisner

NATIONAL GEOGRAPHIC
WASHINGTON, D.C.

Published by the National Geographic Society
1145 17th Street N.W., Washington, D.C. 20036
Copyright © 2010 Jean-Michel Cousteau and Éditions de l'Archipel. All rights reserved. Reproduction
of the whole or any part of the contents without written permission from the publisher is prohibited.
ISBN: 978-1-4262-0683-2

Library of Congress Cataloging-in-Publication Data

Cousteau, Jean-Michel.
  My father, the captain : my life with Jacques Cousteau / by Jean-Michel Cousteau with Daniel Paisner.
    p. cm.
  Published in French under title: Mon père, le commandant.
  ISBN 978-1-4262-0683-2 (alk. paper)
  1. Cousteau, Jacques, 1910-1997. 2. Oceanographers--France--Biography. I. Title.
  GC30.C68C682 2010
  551.46092--dc22
  [B]
                                        2010006114

The National Geographic Society is one of the world's largest nonprofit scientific and educational organizations. Founded in 1888 to "increase and diffuse geographic knowledge," the Society works to inspire
people to care about the planet. It reaches more than 325 million people worldwide each month through
its official journal, *National Geographic,* and other magazines; National Geographic Channel; television
documentaries; music; radio; films; books; DVDs; maps; exhibitions; school publishing programs; inter-
active media; and merchandise. National Geographic has funded more than 9,000 scientific research, conservation and exploration projects and supports an education program combating geographic illiteracy.

For more information, please call 1-800-NGS LINE (647-5463) or write to the following address:

National Geographic Society
1145 17th Street N.W.
Washington, D.C. 20036-4688 U.S.A.

Visit us online at www.nationalgeographic.com

For information about special discounts for bulk purchases, please contact
National Geographic Books Special Sales: ngspecsales@ngs.org

For rights or permissions inquiries, please contact National Geographic Books
Subsidiary Rights: ngbookrights@ngs.org

Interior design: Cameron Zotter

Printed in the United States of America

10/WCPF-CML/1

To my mother, our compass

JYC, wipe your fins before coming in!

(Drawing by Dominique Serafin)

CONTENTS

"The road to paradise is paradise."
—*Spanish proverb*

TWO SALUTES

June 30, 1997. It is a day of sadness and ceremony, of celebration and deep affection. For me, there is also curiosity and wonder. I am standing in the first pew, in front of a coffin draped by the French flag. Next to me is my father's widow, dressed in black. She is flanked by her two children—my half siblings. For the longest time, we did not know each other. Indeed, we hardly knew *of* each other. Together, we look nothing like the family portrait I have carried in my head for the past 59 years. That picture should include my younger brother, Philippe, dead now for 18 years, and my beloved mother, Simone Melchior Cousteau, gone as well for nearly seven years.

This picture? Well, it is an image no one in my immediate family could have envisioned—and yet, because we are

9

assembled here for my father's funeral, it is the one that we now have to consider. It is here before us, and we are in its middle; and I am afraid that it is an image that will endure. I can close my eyes and imagine another more intimate portrait, but eventually this one will push its way into my thinking. Even a man who built his life around the power and majesty of moving pictures must leave this world with a freeze-frame memory; and so, like it or not, this will be my father's legacy.

Jacques Chirac, the President of the French Republic, graces the magnificent cathedral and honors my father's memory with his presence. He sits alone, at the front of the church. Behind President Chirac sit prominent mourners from the disparate worlds of politics, culture, arts, and sciences. Such was the breadth of my father's life that it reached into so many different corners, touched so many different people—many I only know by reputation or by their pictures in the newspaper. My father's peers of the French Academy are also present: Hélène Carrère d'Encausse, Alain Decaux, Bertrand Poirot-Delpech, and many others. There are ministers, too: Dominique Voynet of the Environment; Louis Le Pensec of Agriculture and Fishing. Next to them sit the mayor of Paris and his wife, the chief of the Paris police, the chief of police of the Ile-de-France region, and on and on.

It is a grand and fitting assemblage, but beneath the homilies and stirring tributes, I am moved to consider my father's place in this hallowed hall. He was not the most religious of men. He was born into a traditional Catholic family, but my paternal grandparents were not particularly religious. That said, this is a funeral, a memorial, and so I do not give the question of my father's faith a second thought.

A large crowd is gathered on the steps outside the cathedral—people of all ages, all classes, all endeavors. Through my father's work and passion, they discovered an amazing new world, a place of wonder and beauty that had been out of reach until he found a way to film it and bring it into their homes. Through his eyes, through the testimony of his camera, through his clever inventions and innovations, the sea has become magically alive for these people, and they are here to show their gratitude. That is how I have come to see it. They are here, some of them, donning red woolen caps of the sort my father had famously worn on his expeditions. They are here because, at one time or another, my father—known by the familiar nickname of JYC to friends and family, blending the initials for Jacques-Yves Cousteau in such a way that it came out sounding like "Jeek"—had helped to lift them from their troubles and set them down in a place of mystery, tranquility, hope, and wonder. They are here because he had stirred their adventurous spirit.

I am following the funeral procession, but the sound of the cathedral's big organ draws me back to thoughts of my father. I hear him in the solemn music. I see him in the eyes of the other mourners, many of whom I have never met. He was a man who kept his distance, who had been authoritarian by choice, who was not always warm to even his closest friends and associates. He was a man who believed in efficiency above all. He was a man who would not be denied. But alongside all of this, he was also a man with an undeniable charisma, a man who always achieved his goals— and, primarily, a man of such single-minded determination that he would not give up on a goal until he had achieved it. He dreamed big, and he lived even bigger, if such a thing is possible. He was transported—transformed—by his life's work and so, too, were those who were carried along in his marvelous wake. He sought joy, release, freedom, and wonder. He made his living from these emotions and delivered them in return.

Let us make no mistake. This is a funeral for a man of the sea. The marines who carry his coffin outside the cathedral remind us of this, but the picture is a bit murky. The dots and pixels do not quite come together. As the procession passes beneath a dank, gray sky, I cannot help but wonder at my father's final resting place—in the family vault, in Saint-André-de-Cubzac. Tomorrow, reports of his funeral will appear in newspapers all over the world, but how many of those accounts will consider the absurdity of his exit? A man of the sea, buried on dry land? Such a strange destination for a man who lived his life on the open water. Whose first wife, his companion for nearly a lifetime, had been buried at sea—in the Mediterranean, in front of my parents' home in Monaco, in the deepest trench. Whose younger son, the exemplar of his boundless spirit, was buried at sea as well—in the Atlantic, off the coast of Portugal. When it comes to that, I also intend the sea to be my final resting place. But my father? Our captain? Why bury on land a man who for so many years had been *Calypso's* lover?

Once again, with finality, this was my father. Captain Cousteau. A man of contradictions. A man of many passions I can only try to understand. I think, *Okay, Jean-Michel, you could never really figure him out in life, not completely, so what makes you think you can do any better now that he's gone?*

His coffin passes, and I salute him one last time—or, at least, one last time on this Earth, for there is still the matter of paying my respects to him in the water. This happens soon enough, but it surprises me just the same. A few days later, I am given a chance to say a more personal good-bye; it is a chance I do not see coming, but it washes over me like it is inevitable.

I am back home in California, still shaken by my father's death. The thought will always shake me up, I realize now, but I am still getting used to it, even as the balance of my life and work come calling. It is a fitting call: Some weeks earlier, I had committed to participate in a fish count with a local marine-life organization. "Counts" like this one are conducted all over the world, in much the same way the Audubon Society might conduct bird counts. The idea is to identify all the different species of fish and offer an educated estimate on numbers and perhaps a forecast for future preservation efforts. It's a very basic and useful way to track the degeneration and regeneration of certain areas of the world's oceans, so I am happy to participate—happy, too, that the outing might offer a chance for me to dive and perhaps place my father's death in some kind of context.

If nothing else, I think, it will be a comfortable distraction.

There are 83 other people on board the boat as we pull away from the marina in Ventura, a small town about 30 miles down the coast from Santa Barbara. We sail to an island called Anacapa, the closest island to the mainland, about 16 miles out in the ocean. I am in a pensive mood. I had been emotionally prepared to shoulder the death of my father, it occurs to me. He was an old man. He had been in failing health for the last five months of his life. His death was not sudden. And yet, there is no way to prepare for such a loss. Even with time to get used to the idea, there is a sudden, overwhelming sadness.

The weather is crappy, which dampens my mood even further. I am not good company, among these people, and that makes me sadder still. When he was out on the water, preparing to dive, my father could be one of the most magnanimous, big-hearted people on the planet. His personality, which was already colored with larger-than-life strokes, became bigger still. He would not have understood how a Cousteau could stand on the deck of a ship, preparing to dive, in a despairing mood. And yet here I am, sullen and despairing. I am facing an important task—one I'd been looking forward to, in fact—but I am wishing it away before I have even begun.

The day is overcast and thick with fog—but then, as we approach the island, the skies seem to magically clear. All of a sudden, as if it had been choreographed, the sun begins to shine. The water, which had been choppy on the ride out from Ventura, is now completely calm. It is the strangest, most wonderful thing, but I do not think anything of it at the time. I notice, but that is all.

Once again, it is here before me, and I am in its middle.

I am dressed in my diving gear, ready to jump into the water, but then I have an idea. It alights in my head with no help from me. I turn to the others on board and say, "Ladies and gentleman, I have a small favor to ask. Might I have a few moments by myself, in the water, before the rest of you join me?"

Under different circumstances, this might have appeared an odd request, but everyone on board surely knows about my father's death. Since I was a small boy, people on the water have known me as the oldest son of Jacques-Yves Cousteau. His reputation has always preceded me—in life as it does here now in death. Many of the "fish counters" on board know that this is to be my first dive after Dad's funeral. Everybody seems to understand, and without a word, the crowd that had gathered on the deck silently parts, clearing a path for me. I jump into the water—directly into a very large, very beautiful kelp bed. It offers such a striking sight, as I swim by myself among these enormous, 60-foot seaweed stalks, swaying gently in the swells. I'd always enjoyed swimming in this type of setting, because it reminds me of the birds flying among the treetops in Sequoia National Park. You are as free as you can be. It's lovely, and peaceful, and quite soothing. And so it is on this surprising afternoon. I continue to swim about in this solitary manner, all the time thinking of my father, the great Jacques-Yves Cousteau, the man who opened our eyes to the vast wonders of the sea. He did this not only for the good of all mankind, of course, but also for me. I consider this. It is a deeply personal, profoundly selfish realization, but at just that moment, I hold it close.

Such a gift he has given to me! Such an inheritance!

I am about to swim back toward the boat and motion for the others to join me when I notice an unusual opening in the kelp. All around there is this dense and marvelous kelp forest, which appears to collect around a silky, sandy bottom. The way it appears before me, at just that moment, is almost otherworldly. The sun's rays are shining brightly through the kelp, lighting up this little, inexplicable patch of sand—like a spotlight on an empty stage. All around me there are small garibaldi, brightly colored damselfish native to the area. In normal light, they're a fantastic, shimmering orange, but here, in this bright sunlight, they are like festive candles, a string of party decorations announcing some underwater fiesta.

What can I do but follow these playful, carrot-colored fish to the sandy bottom? What can I do but give myself over to their sweet allure? I drop to my knees and fall on the spot, and I am overcome with emotion. It is almost mystical. I have been thinking of my father, of course, and now it feels as if he is here with me, on this sun-splashed ocean floor.

For the first time since his death, he is near.

That ceremony at Notre Dame cathedral? That was for the cameras, and for all of those French luminaries. It was not for him, and it was most certainly not for me. The burial at Saint-André-de-Cubzac? Another formality, a necessity. But he is here, among these dazzling seaweeds, among these sparkling fish—ah, the ultimate luminaries—on the floor of this impossibly beautiful kelp bed.

This is the real Cousteau, I allow myself to think. This is where he lived and where he will remain.

I am in his world—and he, at last, is in mine.

Beginnings

I have always found it curious that my father's family had almost nothing to do with the sea. It is as though he came to it on his own, like a calling, without the benefit of familiarity, proximity, or custom.

My grandfather Daniel Cousteau was an attorney with a bit of wanderlust in his soul. My grandmother Elizabeth Duranthon Cousteau was a typical Irish housewife. She ruled the household with a firm hand and a warm heart. My grandfather, whom we all called Daddy, worked as a kind of chief of staff—almost like a personal chargé d'affaires— for two prominent businessmen. He worked first in the employ of a British gentleman, and later for an American. Both were millionaires, which was no small distinction in the years leading up to the First World War. Tellingly, both

17

of these tycoons had lavish, expensive yachts, for which my grandfather was often responsible. He would make sure the crew was in place, the wine was appropriately stored, and all the permits were in order.

As my father later told me, those ships were almost always at sea, crisscrossing the Atlantic, but as a young man, my paternal grandfather was often at home in France, on decidedly dry land. Only later, as my father grew up, did Daddy begin to satisfy his passion for exploration and travel. Over the years, he began to visit New York with more and more frequency, eventually importing wine and champagne through a warehouse he operated beneath one of the city bridges. He was away from home for longer and longer periods, but it would be some time before my father joined my grandfather on his travels, before he would taste the spray of salt water in the ripping winds of the open sea. As a child, Jacques-Yves Cousteau lived mostly in the Bordeaux region of France, in Saint-André-de-Cubzac, a small village in the Gironde Province, where he is buried today. As a boy, he lived in a simple home, with no electricity and little to capture his imagination beyond—well, his imagination.

I suppose it was only natural and to a certain degree inevitable that wine played a major role in my father's growing up. He was French, after all, from the Gironde. Wine was in the soil, in the air, and all around. Some might suggest that it coursed through the region like lifeblood. One of my father's great-uncles owned one of the major wineries in the area; unfortunately, he was also a gambler, and he ended up squandering all of his assets on an ill-advised trip to Monaco when my father was still a small boy. The great-uncle lost his vineyard, his properties, even his beloved car—a Bentley, which my father often spoke about in wistful terms. Who knows, if his great-uncle had been a bit more careful with his money, my father might have drifted into the family winemaking business.

As it happened, he merely drifted—and, drifting, he developed a spirit of travel and adventure to match his father's. He was a Cousteau, after all; try as he might, he could not sit still.

My father was a skinny child and not particularly healthy. He suffered from a variety of stomach ailments, including a chronic case of enteritis, which left him feeling weak and out of sorts. He never thought himself a natural or gifted athlete. As a result, perhaps, he developed a fierce determination. What he lacked in strength, he made up for in will. Throughout his life, whenever he set his mind to an idea or a goal, he pushed resolutely toward it. One of his lifelong obsessions was filmmaking, and it offered an early outlet for my father's creativity. He saw himself as an auteur at a time when the exciting wave of French cinema began to capture the world's attention. Somehow, at 12 years old, Dad came into possession of a 9-millimeter Pathé camera. Not an 8-millimeter camera. Not a super 8. Not a 16. How he came to own such an expensive piece of equipment was never made clear. Knowing my father and how he almost always got what he wanted, it's possible that he only had to wish for it to make it so.

Today, most people have never even heard of a 9-millimeter camera, but it was my father's pride and joy throughout his teenage years and into his young adulthood. He carried it with him every-where. In later years, he confided that the camera helped him past a youthful shyness that I could never imagine in the gregarious man I came to know and admire. He said that, with his camera in hand, he could talk to pretty girls, move about in unfamiliar surround-ings, or act bravely or brazenly in situations that might otherwise find him reticent and withdrawn. The camera offered a license to be bold. To listen to Dad's stories of his teenage years, you would not think he ever set that camera aside. Soon, he became quite expert at its inner workings. He was constantly disassembling it to understand how it worked, cleaning it, refashioning it in such a way that it might be more suited to his ever-changing needs.

He used that camera to make many, many films—and, happily, much of this early footage survives. There were other cameras, too. When he wore one out, Dad scraped together some money and purchased another. Before long, he had a whole collection, and the outgrowth of that collection was a mountain of footage. Just before his 75th birthday, I was rummaging through his belongings, hoping to put together a retrospective of his life, which I planned to call "The First 75 Years." My dad thought this was a foolish enterprise. He wasn't interested in the past. (For years, he was not even focused on the future, choosing to live very much in the present.) He was still in relative good health, still diving, still exploring, determined to look only ahead. Despite his protestations, I found some of his 9-millimeter footage in his basement and had it transferred to a 16-millimeter format, and what came back was quite a surprise. Apparently, he'd written these wonderful little melodramas and acted them out with his friends. In one, a fellow and his girlfriend are driving a car. When the fellow pulls over and steps away from the vehicle, my father enters the shot and proceeds to steal the car—with the girlfriend still in the passenger seat. He drives away, but the angry boyfriend gives chase. The stolen car stalls, allowing the boyfriend to catch up, and the fellow pulls my father from the car and drags him to the ground.

In another, my father appears on a small boat with a young woman. The two begin to argue. (For some reason, arguing seemed to be one of Dad's central themes—perhaps he thought it made for a more gripping story line.) There is no sound, but the camera pulls in close on the couple, as their argument intensifies. Finally, my father stands up in the small boat, and the young woman pushes him overboard. It could be said that the surviving shots of my father disappearing into the lake represent the earliest footage of *The Undersea World of Jacques Cousteau.*

In truth, it would be some time before my father became comfortable in the water. He was never an especially strong swimmer. He was just okay—he himself admitted as much. He was comfortable in the water; in fact, he often said he felt most at home there; but he was not a natural. He had to work at it. Throughout his life, he regretted his shortcomings in this area, and yet his determination set him apart and saw him through. For example, at about age ten, he traveled to the United States with my grandfather one summer and was then promptly dispatched to a children's camp in Vermont. One of the primary activities at this camp, it turned out, was horseback riding, but young JYC hated horses. He could not understand those animals, he always said; and, more to the point, they could not understand him. Whenever it was his turn to ride, he refused. Eventually, the camp director came up with another activity for my father as a form of punishment. He had to jump into the lake and remove all the dead branches and debris from the swimming area so that the other children would not get hurt. The branches would gather by the diving dock and present a hazard. It was a difficult assignment, and in future years my father would reminisce about his time in Harvey's Lake, diving beneath the surface, fantasizing about a way to breathe through the hollow reeds that accumulated by the swimming dock so that he could swim underwater while he completed his task. Eventually, he taught himself to do just that—and here again, he seemed to be preparing for a lifetime underwater. Though he could not swim especially well, he could fill his lungs and remain submerged for extended periods.

He was only a mediocre student, a fact that nearly kept him from another of his childhood obsessions—or determinations. For as long as he could remember, JYC had wanted to become a naval officer. He also wanted to be a pilot, and perhaps even a doctor, and underneath all of that, a filmmaker as well, but he believed those dreams would flow from his primary obsession: A naval officer—that would be his first and true calling. He got it in his head early on that he would be a world traveler and that a naval career would be his ticket to exotic ports of call and boundless adventure, only it appeared for a time that his poor record in high school might keep him from enrolling in the naval academy.

In the end, he did enter the academy, l'Ecole Navale, and he even managed to excel in the classroom, graduating with distinction as a gunnery officer. Immediately upon graduation in 1933 he boarded the French helicopter cruiser, the *Jeanne d'Arc*—named for Joan of Arc, a national heroine who had famously repelled the English invasion during the Hundred Years War—joining other young officers on a teaching and training voyage around the world. It was a central part of the young officers' education, a way for these new graduates to get a feel for the planet, a feel for the sea, a feel for other cultures. And, it is a practice the French Navy continues to this day.

My father was plainly thrilled to be aboard the *Jeanne d'Arc*. You can see his youthful enthusiasm on full display in the home movies he shot on his precious 9-millimeter, which he took along with him in his gear bag. Such brio! Such hope! There are images of him and his cohorts laughing and struggling to make themselves understood in a restaurant in Japan. There is "scripted" footage of the freshly minted officers playing out amusing scenarios against every imaginable backdrop. The ship even made its way to California, and there is footage of my father on a studio lot with Douglas Fairbanks and Claudette Colbert, fumbling with a cigarette. In those days, my dad did not smoke, although he did take up the habit later on, but I suppose when a great Hollywood star like

Douglas Fairbanks offers a young French naval officer a cigarette, he is inclined to accept it.

My father was in and out of various ports of call for a full year, after which he was stationed in Shanghai for another year. In Japan, in fact, he met my mother, Simone Melchior. It was just a brief encounter, but it made an impression—on my mother, anyway. Actually, Dad met his future father-in-law first—a distinguished French businessman named Henri Melchior, who was living in Kobe as a kind of de facto ambassador of France—and for the time being, this was the more memorable encounter. Somehow, my father had received an invitation to a reception of some kind at the "ambassador's" home. His daughter Simone was only 17 years old at the time. She had spent a good portion of her childhood in Kobe, a Japanese port in Osaka Bay, and the Japanese influences of her youth would remain for a lifetime—in her style of dress, her tastes in food, her stoic demeanor.

Dad, a young officer of about 26, was mostly interested in meeting Henri Melchior. The man's daughter would have to wait. You see, Henri Melchior had been a frigate commander, now retired from the navy and serving as a director of Aire Liquide, a French company that had become enormously successful selling compressed air. The company's stock is still traded on the Paris Stock Exchange, where it is considered a blue-chip holding. Despite my family's long association with the company, I have always believed its success was fairly symbolic of our free market society. Think of it: The company's core resource and principal product is air—a precious commodity, to be sure, but one that is free and plentiful. Nevertheless, they've developed all these different ways of compressing the air, storing it, selling it, and putting it to all kinds of good and effective uses. To me, that is a clever business model. By now, they've expanded into other gases and new technologies, and you can see evidence of their reach all over the world. In San Francisco, for example, there is a pipeline for wine that is powered by Aire Liquide, so the applications are varied and exciting, but it started in a very basic way—and my maternal grandfather was a part of all that.

Soon, my father would be a part of that as well, but I do not wish to get ahead of my story—of my father's, for that matter—so I'll return to the introduction of my mother. She came from a long line of French admirals. She was the granddaughter of Admiral Jules-Bernard-François Melchior and Admiral Jean Bahème, and the great-granddaughter of Admiral Charles-Joseph Bahème. It was said in her family that she had salt water in her veins, although the seafaring gene appeared to have skipped her two brothers—Maurice, one year older, and her twin brother Michel. My mother's dream was to become an admiral as well, although in those days, a young girl could hardly dream of such a future. Instead, she passed many long hours gazing contemplatively toward the water and staring longingly whenever she came across a naval officer in full uniform, as she did when she first met my father. My goodness, she was mesmerized by those young men, because they represented what she could not have, what she so dearly and desperately wanted—a life at sea. To her, it was the most romantic life she could imagine, and so this first brief encounter with my father kindled something in her that she could not experience on her own.

For his part, my father took scant notice; at least, that is what I was always told, although I can hardly imagine that this was so. My mother was a beautiful young woman, impossible not to notice, so I can only suppose that my father was distracted—a quality I would come to know all too well, for my father would frequently lose himself in thought or wonder or imagination. Also, at 17 my mother was far too young to be the object of any sort of romantic intentions, so perhaps Dad's indifference was age-related as well.

And practical. There was no profit in taking up with a mere girl, after all—not when there were so many perfectly lovely women available to a man of his station.

There is an interesting side note to my father's naval career. Truth be told, it almost didn't come to pass, and he might never have appeared before my mother as a likely suitor. It wasn't his lackluster approach to academics that nearly did him in. And it wasn't his skinny, nonathletic build. It was his heart. During a childhood illness, he had developed a heart murmur. It wasn't considered serious, and it certainly did nothing to prevent him from living a long and active life, but if it had been detected during his military medical examination, he would have never been allowed to join the navy.

In later years, my father would relish in telling and retelling how he managed to conceal his condition from the navy doctors. He knew all about the heart murmur, of course. He knew what it meant in terms of his ability to serve. And yet he stood before the navy doctors with a sense of entitlement and purpose—not quite knowing how he might pass his physical examination but that he surely would.

At one point, he was preparing to blow into a small device to measure his lung capacity, when the doctor was called away on another matter.

"Don't move," the doctor instructed my father. "I'll be right back."

Young JYC did as he was told. He stood stock-still, uncertain of his next move. In several versions of the story, my father was sweating nervously as he awaited the doctor's return. In others, he remained cool, relaxed, unconcerned. In reality, I imagine his true character fell somewhere between anxiety and calm. After a long while, the doctor returned. He approached my father, who had been stripped to the waist.

"Where were we?" the doctor asked, reaching for his stethoscope. "I've already examined you, yes?"

My father did not think it was his place to disagree with the doctor on this, and in this benignly duplicitous way, he took his place in the academy. He simply nodded in ascent as the doctor shouted, "Next!" Then he buttoned up his shirt without a word.

Such was the good fortune that attached to my father—like a charm bracelet.

My mother was fairly smitten by the future Captain Cousteau, who was nine years her senior, although it would be some time before my father had an opportunity to return her attentions. In the meantime, he decided on something of a whim to pursue another of his lifelong passions and put in for a transfer to the Naval Aviation Corps. In practice, it was only a slight change in his career plans—one he believed the navy could readily accommodate and that he could certainly achieve. He went to the aviation center of Hourtin, in the Gironde, to begin his pilot training. By his own estimate, he would complete his training in six months, so he set about achieving his goal with his customary determination and focus. He would not be kept from his dreams. Rather, he would collect them one by one, with his 9-millimeter camera in hand to record his every move. And for this adventure, he would add still photography to his list of passions and talents, bringing back hundreds of dazzling aerial photographs he took during his first flights.

But fate put an end to these particular hopes and dreams for my father, in what was to be the first consequential disappointment of his young life. After six months of training, his goal within reach, he lost control of his car on a rainy night in the Vosges Mountains. Inexplicably, his headlights went out as he approached a treacherous bend in the road. Years later, my father surmised that he must have reached to the dashboard to turn on his high beams and inadvertently turned off his headlights instead. In any case, he lost control of the vehicle and ended up in a ditch. Mercifully, there were no other cars on the road, so he was spared a head-on collision and no one else was hurt in the crash, but that same bit of mercy was also costly, because there were no passing drivers to offer help. He was marooned—shipwrecked—on land. My poor father could never remember how long he lay there, helplessly, before he eventually had to drag himself from the wreckage and walk several miles to the nearest house. As he recalled, it was more like crawling than walking, although even this was especially painful. One arm was shattered; the other one was in only slightly better shape. When JYC finally made it to the local hospital, the doctors determined that he had suffered 27 fractures. In this context, it is a clear indication of the extent of JYC's injuries. A lesser man—a different man—might have retreated within himself after such an accident, but my father could only persevere. It was the only way he knew—forward! And so he pressed on.

For a day or two, while he was recuperating in the hospital, it was feared that the doctors might have to amputate my father's shattered arm. Gangrene had set in and would likely spread. But Dad refused to allow the surgery. He said, "I would rather die than live a life without both my arms." The prospect went against his nature. He would not accept the doctors' diagnosis, and he steadfastly began a series of painful, arduous exercises that would eventually lead to an all but full recovery. The accident would leave him with a depression in his chest that he would never lose, and he would never again have the full use or range of motion of his left arm. As a result, of course, his dreams of becoming a navy pilot were also shattered that night on a wet mountain road. He would have to find a new purpose, a new outlet for his passion, his enthusiasm, his determination.

And yet, just as this car accident closed a door on one of my father's long-held dreams, it opened a window on an opportunity he had yet to consider—and, it led in an indirect way to another chance encounter with my mother. When he was discharged from the hospital, he was transferred to the naval base in Toulon, in the south of France on the Mediterranean coast, where he immediately began an intensive course of physical therapy. As part of his pre-scribed regimen, he began spending a great deal of time in the sea, strengthening his damaged hand and rediscovering the use of his arms. For years afterward, he would speak of this time in the waters off Toulon as elemental, because it also strengthened his skills as a swimmer and ignited a new passion for the sea.

There, in the harbor of Toulon, splashing among the navy ships and recreational vessels, my father had a revelation. It would change his life, he always said. He had borrowed a home-made diving mask from one of his friends and soon developed the habit of diving deep below the surface, then clambering his way back up as a form of exercise. The resistance of the water offered the perfect environment for his rehabilitation, and as he broke back above the surface one afternoon, he paused in such a way that his mask remained half-submerged. For a brief moment, through the top half of his mask, he could look out into the crisp, bright sunshine and see the heat of the day shimmering on the cars passing along the pier. In that same brief moment, he could see dozens and dozens of fish swimming haphazardly in the sea. He saw each scene like a filmmaker, a photographer, and marveled that there could be two distinct, disparate worlds unfolding before him in the same frame, at the same time. It was a boundless, wondrous view.

It lit something in him, this revelation. Indeed, I often heard him speak of this as a kind of life-changing moment. Certainly, it altered his point of view. He started to think about this whole world beneath the surface of the water, a silent world that went largely unnoticed, in a way he never had before. He wanted to see under the keel of his ship. All of a sudden, this became a great frustration for him, that he could not see beneath his ship. A frustration and a fascination, both. He wanted to see what wonders lay in wait on the ocean floor. He wanted to swim like a fish. He wanted to live fully in each of these worlds, above and below, all at once.

At the time, my father did not think anything could be done with this revelation, but it was a revelation just the same. It became another one of his obsessions. He set it aside, but not quite. The idea stayed with him. He tinkered with it, but only in his head. He continued to dive and swim and strengthen his arms, and whenever he broke back to the surface, he tried to re-create that moment when these two worlds were framed for him in such an impossibly beautiful way. The juxtaposition was a constant puzzle. But that was where his fascination would remain, he believed, because there was no place else to put it. So he continued in his work on the base and redoubled his efforts to regain the use of his arms and his hand. Life would go on for him as before, he determined. He would be whole once more—physically whole—only now there would also be this strange emptiness welling within him. He considered it in this way: Following his car accident, all these pieces were missing in his foundation, and now as he struggled to put himself back together, there were all these new pieces to consider. There was this idea that he was missing something profound and meaningful just below the surface of the sea—something he could never really touch and never fully know.

During this time, my father visited Paris and found himself at a social gathering with other naval officers. It was to be an evening out with his companions, a welcome chance to unwind away from the base—to discuss developments regarding the signs of impending war in Europe and how the developments might apply to them. Mostly, though, it was an opportunity to meet some of the more eligible, more elegant young women of polite society. It might have been a time of political upheaval—but these were young men. They had their priorities.

There, on the other side of the dance floor, was my mother. She was now about a year older than she had been at the time of my parents' first meeting—and, naturally, so was he. Much had happened in the intervening months. My mother had gone from girlhood to maturity. My father's worldview had undergone a great change as well, and this time he took special notice. It was another revelation, I suppose, and he crossed the dance floor and asked the lovely Simone Melchior to dance. Here again, looking at the scene like a filmmaker, it is possible that my father had framed the shot in such a way that he could see two worlds at once. There was my mother, looking young and beautiful, and there was the world he knew, in and around the French Navy. Here, though, both worlds were within reach—and so, my father did what he did best. He reached. And, they danced.

They were quite taken with each other, I was told. And, by all accounts, it was a suitable pairing. From time to time, my mother would admit to me that if she could not be a naval officer herself, then the next best thing would have been to marry a naval officer, and in this way, enjoy the same life by association. But I do not believe there was an ulterior motive in her pursuit of my father. And neither was there motive in my father's pursuit of her, despite what some have written in regard to my grandfather's position with Aire Liquide and JYC's future pursuits in that area. Certainly, the fit was good and natural on a great many levels, and there is no denying that my parents seemed to complement each other in a variety of ways. But most of all, they were drawn to each other, and soon enough they were transparently in love.

To be sure, they found a reason to get together the very next day, most likely without the knowledge of my maternal grandparents, who were no doubt keeping careful tabs on their teenage daughter. Understand, I have no reason to suspect that they would not have approved of a courtship with my father, a well-respected naval officer, but I was always told that my parents' first several dates as a couple were conducted surreptitiously. Apparently, the rarified air of the Melchior residence on the Place des Ternes was considered a bit beyond the reach of the young Jacques-Yves Cousteau, who had yet to make a name or a reputation for himself.

That would soon change, of course.

After a whirlwind courtship, my parents were married in the church Saint-Louis-des-Invalides in Paris—a great honor for a naval officer. They had a marvelous ceremony, which included my father's fellow sailors in formal dress, making a steel archway with their swords for the young couple. It was all very, very official—and very, very grand. My father was dressed in his naval whites, and my mother in a gorgeous dress, looking very beautiful.

It was to be the beginning of a long, eventful marriage.

My Arrival

I was born on May 6, 1938, a little more than nine
months after my parents' wedding. The story my father
always told was that I was conceived in the sleeping car
of the train that carried the newlywed couple to Italy for
their honeymoon. I don't know that this is true, although
my father liked to tell the story with great embellishment,
suggestively mimicking the rickety-rack sounds and movements of a passenger train, invoking for his listeners the
passion of two young lovers on their wedding night.

My father was an animated storyteller, and here without
the help of his precious camera, his material appeared to
match his gifts, as he quite literally gave life to the story of
how I was given life and how the life he imagined for himself and my mother might begin. I have chosen to believe

the story for its poetry, for its symmetry—and, for what it represents. In a very real and tangible way, my mother's pregnancy marked the formal beginning of my parents' lives as a couple, as a family. Prior to that, they had merely been caught in the magnetic pull of two forceful personalities—in love with the idea of marriage, I am guessing, and the conventional trappings of adulthood as much as they were with each other. Do not misunderstand: They were drawn to each other, this much was clear, but it remained unclear what kind of picture they might make over time. Their relationship was still taking hold. As common ground, they had the sea itself, and the hardly coincidental fact that both were the children of itinerant fathers, but they would need more to their union than that. They had only been married for a short while; there was some haste to their courtship and engagement, as well, which meant they could not have known each other very well. Without a child, they were simply two young lovers, embarking on a journey together—anxious to get started and filled with hope and possibility, but tentative and uncertain at the same time. As parents, suddenly, their partnership represented so much more, and all of that hope and possibility would finally take shape, while that uncertain picture would come into focus.

Throughout the winter of 1938 and into the spring, some measure of my parents' routines and their personality as a couple began to fall into place, even before I was born. Lieutenant Cousteau and his pregnant bride settled on the outskirts of Toulon, a city of contrasts, a Mediterranean seaport that could appear dirty and neglected in one respect, and lively and luminous in another. My father fit himself quite happily in each of those contrasting spheres, as he would seek to join or coexist within disparate worlds throughout his life. Looking back, I suppose this was an early extension of the revelation he experienced through the two halves of his borrowed diving mask, taking in the worlds above and below. Here, he would move among the privileged few and the not-so-privileged many, as if he belonged no place else. This dichotomy reflected one of my father's great gifts, to ease into almost any setting, primarily because he never saw a situation or circumstance in which he did not belong. It was simply who he was—a man of the world, a man of all worlds.

He had emerged from behind his camera and presented himself as a man who could make himself comfortable as he made those around him comfortable as well.

Were it not for the presence of the naval base in Toulon, I imagine the town's prospects might have been more in line with those of any other coastal town in the south of France. But the comings and goings of so many sailors, together with their wives and children and a great many dignitaries who would regularly cross the sea to our shores, gave Toulon a kind of bustling charm. There was always some excitement or other. There was the working class of the seaport mixed with the more privileged classes of naval officers and visiting merchants, and the resulting soup was a delightful blend of rogues and royalty, lowbrow and highbrow, and everything in between.

This contrast was most readily apparent in my father's pursuits. He did not have a great deal of free time, but he was determined to enjoy his precious days off to the fullest. As an officer, he was entitled to swim at La Royale, a kind of officers' club complete with a well-appointed bar and an elegant restaurant. As the name suggested, members were treated like royalty. La Royale was a favorite off-duty gathering place for my father and his fellow officers, who would sink into the big leather armchairs in the lounge and share their tall tales of adventure at sea, along with their speculations about adventures to come. In my father's case, however, he mostly enjoyed the swimming privileges that came with his membership, allowing him to traverse the waters off Sanary, in the district of Le Mourillon, which he much preferred to the public beaches in the surrounding areas—where the surfeit of bathers scared the fish my father so longed to observe.

Another favorite gathering place, however, was much more suited to the rank and file of the region's lower classes. La Corniche was a popular nightclub—or boîte, as such a place would be known in those days. Here, as elsewhere, my father felt quite at home with all manner of men and women from all segments of society. My mother would often join him at La Corniche just after they were married, and together they would fit themselves into the scene, although I suspect my father did not fit himself in so easily now that he was married. The reason for this was straightforward: Before taking up with my mother, he considered himself something of a ladies' man, and there was evidence of many past dalliances among the patrons at La Corniche—and so it was not the fit itself that made my father uneasy, but the consequences of that fit. Very often, a charming young lady would approach—now as the very proper wife of an admiral or a medical officer, but once and forever in my father's memory as some Lola or Zaza who had shared an evening at his side.

In later years, my mother would remark quite casually that she could always tell when my father was introducing her to a former girlfriend. There would be a certain glimmer in his eye, she'd say, or a sly smile she would come to recognize as a telltale sign, but she would not begrudge him his past affairs, just as she could not keep him from his old haunts. To his credit, my father never advertised these past relationships to his new wife—but, even so, she could not help but take notice, and underneath her noticing, she found reason to hope that my father's days as a ladies' man had come to an end.

It was in the water, though, that my father began to discover his true self, to shed his past, and to move forcefully toward his future. This happened almost on its own momentum. His regimen of physical therapy continued—and, along with it, his thirst to explore the vast, hidden wonders beneath the surface of the water. Much of his free time was spent swimming with long strokes from the fortress of Toulon toward the open sea. He often swam alone, but just as often in the company of his friends and fellow officers. The other swimmers would only reach to the pontoon, where they would stop to catch their breath. But my father was always eager to keep going—ever anxious, he always said, to recapture a moment of discovery he experienced on one of his dives to the Mourillon seabed a summer or two before. It was a moment he wrote about in his first book, *The Silent World,* and I'll let him tell it here:

> The sea was merely a salty obstacle that burned my eyes. I was astounded by what I saw in the shallow shingle at Le Mourillon, rocks covered with green, brown, and silver forests of algae and fishes unknown to me, swimming in crystalline water. . . . Sometimes we are lucky enough to know that our lives have been changed, to discard the old, embrace the new, and run headlong down an immutable course. It happened to me at Le Mourillon on that summer's day, when my eyes were opened on the sea.

Yes, my father's eyes opened onto the sea, just as my mother's eyes now opened onto the life unfurling before her. Together, they ran headlong toward what they both believed to be a bright and meaningful future. Surely, it was a thrilling time for them. Very early on—while she was still pregnant with me, I believe—my mother came to recognize my father's mission and to embrace it as her own. This happened slowly at first, but their shared passion for the living sea only deepened over time. My mother was still just a teenager, still adjusting to the realization that the man she had married was a man of the world. But all she had ever wanted was a lifetime at sea, and here she was about to experience just that. It was all very nearly close enough to reach, close enough to taste. However, she would only know my father's "silent world" in a once-removed sort of way, tethered to hearth and home because of her pregnancy—and, after that, because she had to care for an infant. She had thrown in with a purposeful young officer who was able to articulate the passion she felt for the oceans and the adventures she longed to experience. In this respect, at least, they were an even match. Theirs was to be a shared mission, she always said—and she recognized this early on. Together, they would help the people of the world understand how connected we are to our vast oceans. Already, that mission was coming into focus, even though it would be some time before either of my parents could conceive of the adventures that lay in wait.

My goodness, they could hardly conceive of me. I was something of a surprise, I was often told—a welcome surprise, but a surprise just the same. Indeed, my mother was nearly still a child herself when I arrived on the scene. The story of my birth has passed into family lore, although I never sought to verify the unlikely sequence of events as it was passed along to me over the years. Still, I much prefer my parents' version to any deeper, fuller truth that might have emerged from a deeper, fuller inquiry. Actually, it is mostly my father's version of events, because the storytelling usually fell to him. For her part, my mother was frequently content to let his recounting stand as a matter of family record. And so, in my father's version, my mother lay back on the kitchen table after her water broke, and my arrival was imminent. My father scurried about, somewhat frantically, collecting and sharpening the big kitchen knives within reach—for what purpose, I could hardly imagine, and he could hardly articulate. Thankfully, they had a midwife of sorts to help my mother through the birth—and, to gently remove my father from the scene. But then he returned to the kitchen when he heard my first cries, dressed now in the fine splendor of his uniform, ready to celebrate my entry into the world.

Once more, to be clear, I cannot be certain that this scene unfolded in quite this way. Nevertheless, this is the picture that I keep in mind and hold dear: my father, in the pristine whites of a La Royale officer, his ceremonial sword sheathed at his side, bursting into the kitchen with a bottle of champagne to congratulate his wife and greet his newborn son.

As moments go, it was surely grand—and, surely French, or so my father always said.

I was born in a time of relative peace. France's involvement in the war did not begin until the following year, and so my father's duties kept him mostly in Toulon, for the time being at least. From time to time, he would leave on a ship for several weeks, only to return to the naval base a short while later, where he would somehow find time to resume his extracurricular activities. These personal pursuits began to occupy more of his attention. In some ways, he later said, his work as a gunnery officer became routine; even as he sailed and experienced great things, his mind kept wandering to his home in port, to his young family, to the wonders awaiting him below the surface of the water. His time in the sea, diving beneath the surface, imagining the world that was waiting for him there, began to occupy his full attention in such a way that it was nearly a preoccupation.

In the summer after I was born, my parents would spend long days at the beach in the company of other officers and their families, diving and splashing and swimming. During these idyllic afternoons on the shore, I was always told, my mother developed a fascination for the ocean to match my father's. It had always been a part of her, but now she could finally dive in—literally and figuratively and every which way in between. Dad taught her to dive deep below the surface and to swim among the fishes, her eyes open to the wonder all around. My mother was a much more natural athlete than my father, and whereas she could not match him in terms of physical strength, she could certainly swim and dive with more enthusiasm than the wives of the other officers. Knowing my father, I imagine this was a great source of pride for him—but it also presented a flicker of opportunity, because he knew his pursuits would only take him as far as my mother was willing to go.

One of Dad's great companions during this period was another naval officer named Philippe Tailliez, who was a few years older (and, therefore, a few years more worldly) than my father. Tailliez, in turn, had a friend named Frédéric Dumas, a well-known spear fisherman in the area. Dumas, who was known among his cohorts as Didi, was not in the navy, but he worked at the naval base in Toulon in a civilian capacity, so the three quickly became inseparable. They took to calling themselves les mousquemers— or, the three musketeers of the sea. It was Tailliez who had loaned my father the primitive diving mask, fashioned from an old pair of aviator goggles, that my father was wearing on the afternoon of his revelation dive in these same waters, and it was Dumas who would indirectly inspire my father to make his dream of swimming among the fishes and truly exploring the vast undersea world in the waters off Toulon an achievable reality.

This happened first as a matter of ingenuity, born of necessity. Well, perhaps *necessity* is too forceful a word in this context, because certainly there was no perceived or pressing need for developments in this area. It's not as if these men were developing new technologies to gain any sort of tactical military advantage or for reasons even remotely related to survival. It was simply a means to a desired end—a necessity of their own making. To be sure, the French Navy would soon make great tactical use of my father's discoveries and inventions, and those of his colleagues and peers, but that would happen later—after Dad and his mousquemers had gotten what they wanted out of the effort.

You see, Dumas was in the habit of donning a homemade mask and primitive fins and diving to the bottom of the sea in search of grouper. He was quite famously capable of holding his breath for as long as three minutes, and after each dive he would reemerge at the surface with great quantities of fish, which he had caught with his spear. At the end of the day, he would return to shore with 80, 100—sometimes 150 pounds of fish. Such an impressive display. Once, on a bet, he bagged more than 200 pounds—in a single morning. My father learned to fish in this way as well, but he could not keep up with Dumas. Tailliez, too, was no match for their friend, and soon my father became more interested in filming Dumas on his deep dives than in joining him as a fellow fisherman. He was content to observe and to record; only, here too, he soon grew frustrated because he could not keep up.

Necessity, no. Ingenuity, yes.

My father was an inventive, resourceful fellow. This aspect of character was at the root of his many gifts. He was still in the habit of disassembling his expensive photographic equipment, reimagining its inner workings, and making the necessary improvements to get the gear to perform in the ways he wanted. He had somehow rigged a housing compartment to allow him to submerge his film camera underwater, for the sole purpose of recording Didi's great

41

dives and, in the process, collecting some of the first breathtaking moving pictures of underwater life. This was his nature. He would not take the word *no* for an answer. It was unacceptable to him. The mere fact that something had never been done before did not mean it could not be done in the future. And so, when he could not find the equipment he needed to do what he wanted to do, he set about creating it. He wanted to be able to film underwater, so he made it happen. Working in collaboration with another friend—Leon Vèche, a gunsmith by trade—he constructed a housing compartment made of metal. In those days, there was no plastic, so the two men built a heavy, clunky case of metal with a porthole made of optical-quality glass, and they further waterproofed the contraption with rubber fittings, hoping it would do the trick.

The initial prototypes invariably leaked. Instead of trying these prototypes out on his cameras, Dad experimented first with inanimate objects, such as stones, to determine the efficiency of his contraption. It would not do to submerge an expensive piece of equipment into the Mediterranean Sea, because even a small leak in the housing compartment would cause irreparable damage. His camera was expensive. The film was expensive. Indeed, everything was expensive in the years just before the war, and my father did not have a lot of money. His meager salary only stretched so far.

Later, during the war, when film stock was difficult to find and prohibitively expensive, this became a small source of tension between my parents and a constant source of worry for my father. And yet, despite the tension, my mother worked tirelessly alongside my father when he stumbled on an ingenious solution to this particular difficulty. One memory stands in illustration. By some divine miracle, he always thought, Dad found an old 35-millimeter Kinamo film camera in a secondhand store in Marseille; very quickly, it became one of his prized possessions and a favorite for the way it captured moving images in a vibrant, vivid manner. In many respects, he preferred it to his reliable 9-millimeter camera. However, during the war, the German army was requisitioning almost all of the 35-millimeter film stock in Europe, for use in reconnaissance missions; not only was the film expensive, but now it was nearly impossible to find. This presented a problem for my determined father, who searched diligently in film shops all over France. Whatever the cost, he was determined to pay it. Wherever he happened to be, he found time to look for his precious film, until one day, he realized that he did not need movie film at all. He saw that any 35-millimeter film would work equally well, so he purchased rolls and rolls of film intended for still photography. Then, he and my mother would stay up half the night, under the bedcovers, gluing the film together 36-frame strip by 36-frame strip, so that it would be suitable for moving pictures.

I have a clear, firsthand memory of this. Other small children might have stumbled upon their parents pursuing some other, more amorous activity under the bedcovers. But in the Cousteau household, there was only my mother and father, jury-rigging a roll of film, painstakingly adhering one strip to the next, careful to keep any light from getting under the blankets, which would have exposed the film and spoiled their hard work.

My first vivid memory of my father is from our house just outside Toulon when I was about two years old—just a few months before the birth of my brother, Philippe. For some reason, I remember that Philippe had not yet arrived, because the house would move to a slightly different rhythm with a newborn underfoot. Even as a small child myself, I could recognize the distinction. Here, as I recall, we were still moving to the beat of an only child; for good or ill, I was at the center of my parents' attentions—if only for the time being. JYC came home from the base one afternoon, dressed in his white uniform, and stopped to play in the garden with me and my pet duck, Gédéon. For a short time, this became our afternoon routine. I remember it well. Oh, how I loved that little duck. He was white, with a yellow-orange beak, and I played with him the way other children played with their dog or their cat. He lived in our house, like a family pet, and my father would join us before going inside, playing and playing. And oh, how I loved being on the receiving end of my father's attention, even if I had to share the attention with Gédéon. He was such a busy man, my father, always coming and going, always plotting his next adventure, that I remember this time with him as precious and special. I would see him approaching in his handsome navy whites, sometimes with his ceremonial sword slung through its holster at his hip, and I would think of him with sweep and purpose and grandeur. I was a small boy, to be sure, but Dad loomed before me in a larger-than-life sort of way, especially when he was in uniform, as he was on this afternoon in our garden. It is an early memory, to be sure, but at the same time, it is one of the last memories I have of a time when I had my father all to myself.

Back then, the custom was for young naval officers and their wives to entertain in their homes, and my mother proved herself a very capable hostess. She had learned by observing her mother, Marguerite Melchior, whom we all called Guitte. My mother remembered full well the excitement that would envelop her home on the evenings when her parents entertained, an excitement she now sought to duplicate in our far more modest home. Unlike her mother, my mother had no staff to help with the preparations. Also unlike her mother, most of my parents' guests were men— young, unmarried officers, as well as married men whose wives felt duty-bound to remain at home with the children. This was a serendipitous turn, as far as my mother was concerned, because she much preferred the company of men, so much so that she often surprised her guests and my father's fellow officers with how easily she could lapse into the salty language of the sea, proving herself time and again to be the perfect companion for Jacques-Yves Cousteau—a beautiful woman who loved to laugh, entertain, and speak like a sailor. As such, she was well suited to the surroundings that would find her soon enough.

My brother Philippe was born a short time later—on December 30, 1940—although our circumstances were about to change by this time. For Philippe's birth, family lore has it that he was born on the same kitchen table; however, it was no longer the same kitchen, and, this new kitchen would not be our kitchen for long. The war was encroaching on the south of France, and it now appeared that the Germans were moving into the region. As a primary naval base, Toulon was a transparent target, so my parents were understandably uneasy about remaining nearby. As a precaution, they decided to move my mother, Philippe, and me to the mountains, where it was thought we would be protected from a German invasion. My father found a little house to rent outside of Megève, across from Italy in the Alps, not far from the Mont Blanc Ski Resort, and off we went with my paternal grandmother, and two young cousins—Jean-Pierre and his sister Françoise. My father stayed behind to fulfill his duty. The curious and telling detail here is that money was tight; in addition to the rent in the Alps, my father's meager salary had to be stretched to cover payments in Toulon as well. He sent us what he could, but there wasn't always money for food.

Paradoxically, my parents considered Philippe's birth a kind of godsend—and I suppose it was. Food was expensive, and in some respects hard to find, but because my mother was breast-feeding her newborn, they at least had a ready supply of breast milk. Of course, this meant that she herself needed to maintain a balanced and healthy diet to produce enough milk, but she managed to do so. For years and years, I would hear stories of how my mother nursed the two of us—one on each knee, one on each breast. We were so hungry, she always said, that we ate greedily, and there was no room in our circumstance for me to be jealous of my baby brother.

In the mountains of Megève, we lived on the second floor of a nice, comfortable house. There were no luxuries to speak of, and even some of our necessities were scarce, but to a small child, we seemed to get along quite well at least. The struggles or hardships my family might have experienced during this period were not necessarily experienced by me and my brother in quite the same way—and in many ways we experienced them hardly at all. I remember enjoying this special time with my cousins, especially Jean-Pierre, who was very close in age to me and therefore a well-suited playmate. As far as I knew, we had everything we needed, although in truth, we didn't really have enough money for food or winter clothing. We never had any bread, for example. And my cousins rarely had any milk. We were friendly with our neighbors, who lived just up the hill, and they kindly shared some of their food with us. They had a small farm, so they had a lot of chickens—and, therefore, a lot of eggs. They also had a ready supply of potatoes.

One of my chores was to run up the hill with my cousin to collect some eggs and potatoes and perhaps a little bit of milk, but we didn't always do such a careful job of it. Once, we ran down the hill, enjoying some game or other, and I tripped and fell. All the eggs were broken, the milk spilled. Even at four years old, I knew we could not go back to our neighbors and ask for a second helping. We would have to do without, until the next time. Jean-Pierre and I managed to salvage the potatoes, but we got into a lot of trouble. Regrettably, I received most of the blame, because I was the one who had dropped all of that precious food. My behavior had cost our family. In those days, parents spanked their children when they misbehaved, and here was my turn to receive a beating. I do not recall whether my mother administered the spanking or if my father was visiting with us that day—but in either case, a spanking was a spanking. Over the years, I collected my fair share from each parent, and the punishment was well deserved each time.

Very quickly, we became a part of the community. My mother made friends easily. It was one of her many gifts, and it served her well over the years, as she tracked the comings and goings of dozens of our diving and film crewmembers aboard ship. She had a charismatic personality and a deep affection for people, and these traits were on full display during our time in Megève. We didn't know anybody when we arrived, but it was not long before she befriended our neighbors up the hill. Soon, she had an entire collection of friends, friendships she would continue to nurture for the rest of her life. In Megève, in fact, she first met Marcel and Gaby Ichac—two lifelong friends who would go on to become great collaborators as well. Marcel was a noted French alpinist, photographer, and film director, who later introduced my father to Louis Malle, who would go on to direct *The Silent World,* Dad's first film, earning both Louis and JYC a prestigious Academy Award. Marcel also joined my father on many expeditions over the years, though it was my mother who initially forged the friendship.

My mother was something of a social animal. It would have been against her character to hole up in the mountains with her mother-in-law, her two small sons, and her niece and nephew without reaching out to other people. She longed for good and interesting company, because she only saw my father infrequently during this period. He came to us when he could. He sent money when he could. And, he sent word when he could—which unfortunately meant hardly at all, because communication was virtually nonexistent.

By now, the war in Europe was on full display. Dad remained stationed in Toulon, although he was going back and forth to Paris, as well. His father was there, as were my grandparents on my mother's side. Later, I would learn that he was also finding time on these trips to Paris to work on some of his "preoccupations." His time was not his own; he could not come and go as he pleased. Because of this, his arrival to our home in the Alps was always unannounced. I remember seeing him one afternoon, cresting over the hill in front of our house on a bicycle. I was three or four years old, and it struck me as a surreal vision. For one thing, I did not expect to see my father approaching, and so I did not trust my eyes at first. For another, I did not expect to see him on a bicycle. Even today, the picture is somewhat incongruous, because he rode all the way from Paris—more than 600 miles. It strikes me now as an incredible distance, but it was typical of my father. Where others saw difficulty and impossibility, he saw solution and opportunity. Travel between Paris and Megève was indeed difficult during the war, but not on a bicycle, not if you had the fortitude for the ride. The distance did not put off a man like my father. He was young and strong and determined. Even the rolling hills and the steep climb into the mountains were no intimidation.

As he drew near, I could see that he had fixed two satchels to his bicycle. In one, he carried a few bags of sugar—a staple he knew we had been doing without. In the other, he carried several film canisters, the master footage of a picture he was working on with his friend Frédéric Dumas, who in addition to his diving and fishing skills had become my father's closest collaborator in filmmaking. Indeed, Didi was a fixture in my father's life from the moment they met, and he was always available to collaborate on any project or adventure. In later years, I'd hear Dad tease his good friend constantly about never having a proper job, and I suppose this was true; Dumas made his living from his largesse, his resourcefulness, and his many physical and intellectual gifts. Don't misunderstand: He made a valid and valuable contribution to every enterprise, but he could not be defined or labeled as any one thing. At times, he would stay with us for extended periods—presumably when money was tight and his personal finances would not stretch for a home of his own.

Dad's plan was to continue on to Toulon after visiting with us in the Alps. He would leave the sugar with us and carry the film the rest of the way to Toulon, where he hoped to edit it and perhaps develop a narration and find a distributor. Didi would most likely meet him there and help him see to the arrangements, just as he always managed to turn up—like a lucky penny, Dad always said.

Our life in Megève was marked by another clear memory. It is the very last memory I have of our brief time in the mountains. One afternoon, when my cousin and I were still about four years old, we were outside playing when we noticed several men falling out of the sky in parachutes—Germans, it turned out. Philippe was with us, too. He had just learned to walk, so that would place the memory in early 1942, perhaps in the spring. I have since tried to understand the sudden appearance of these German parachutists in some kind of historical context, alongside a corresponding invasion or action undertaken in the region, but I have been unable to do so. For whatever reason, history did not record this exchange. And yet, there they were, alighting in our midst like particles of sunlight. We were momentarily confused when we first saw them, because we had no frame of reference for such a sight, but as these paratroopers approached the ground, it became clear to us children that they were merely men. French, German, Italian—we could not tell one nationality from another, although of course Jean-Pierre and I could not understand a word these men were saying. We spoke English quite often in our household, but our facility with language did not extend to German. Even so, we could not tell friend from foe. We could, however, distinguish adults from children, and these men were hardly more than boys. The young Germans passed right by our house, and quite a few of them stopped to notice us and offer a playful wave or a smile in greeting. They were very nice, actually. Later, I would hear stories of German soldiers scaring the French villagers and behaving menacingly as they made their way across Europe, sometimes with their rifles drawn. But these Germans were perfectly pleasant. They fell from the sky like it was the most natural thing in the world, and they waved cheerfully as they passed—and we continued with our playing.

This was my first direct encounter with the war, but it would not be my last. Soon after, we moved to an apartment in Marseille. By this point, I was a bit older and more attuned to outside events—still a child, but a child of the war and therefore accustomed to moments of significance. I listened in as my parents and their friends spoke of the war and its developments. It was a primary topic in our household, but it shared prominence with another pressing matter: the sea. In any case, I learned to draw a line from what I was hearing in hushed tones to what I was seeing from stolen glances. From our sixth-floor window, it was possible to hear the sirens outside, warning of a potential bombing. I could look through the window and see the airplanes. Usually, my mother would come to collect me and Philippe in her arms and whisk us away from the window, so we would not be frightened by the scene unfolding outside. But she could not always protect me from myself. Every now and then, I'd remain at the window long enough to witness the full horror of the war—or, at least, the horror within my limited view. Once, I even saw people being corralled in the streets, lined up against a wall, and shot. There was nothing vague or unclear about what I was seeing, and there was no explaining it away.

I was only a child, but I understood.

By this time, my father had been transferred to the navy reserves. He was no longer considered an active-duty officer, and worked instead in counterintelligence. My mother had some idea what he was doing during this period, but not a lot. He was not supposed to talk about it. My brother and I knew even less. We knew only that he was doing important work and that he still could not be with us all the time. In later years, we learned that he was frustrated in his attempts to get his superiors to listen to his warnings about the threat of the German offensive. He knew, for example, that the French fleet stationed in Toulon was vulnerable to attack. It was inevitable, he believed. He gave several warnings, he later told us, but the naval command refused to act on his intelligence—until it was too late. Finally, when it was apparent to all that the Germans were about to seek possession of our ships, the navy reacted in time to scuttle the fleet. It was their only recourse to prevent the Germans from taking possession and gaining an even greater advantage in the conflict, but it must have been a particular heartbreak to a seafaring man like my father to stand by helplessly while his friends and colleagues sank the very ships that had introduced him to a life on the high seas. His own ship, *Dupleix,* was among the many destroyed. The famous *Strasbourg* and *Provence* battleships were also destroyed, and beneath the destruction I could only imagine my father's despair. His devastation was plain for me and my brother to see, and we all felt the devastation of the harbor, which seemed to burn for days and days as the fires were fed by the fuel spilling from the wrecked battleships, destroyers, tankers, tugboats, and minesweepers.

Much has been written about Lieutenant Cousteau's work in the service of his country. Indeed, some years later, in 1946, he was named *Chevalier de la Légion d'honneur pour les eminents services rendus pendant la guerre*—Knight of the Legion of Honor for eminent services rendered during the war—so his efforts were duly noted at the very highest levels. He himself wrote a great deal about that time in his life, and historians have long sought to document his contributions in this area, but I am mostly concerned with what I remember of him as a father during this period. I do not wish to dwell on matters that I did not experience firsthand with my father—or, at least, on matters I could not yet consider from a more adult perspective. When I think back on those war years, my focus remains on our time together, as father and son.

For example, it was always a time of great joy when he was with us in Marseille—a bicycle ride of only 37 miles from his base in Toulon—but it was also a time of great tumult. Our apartment was so small that we were all very nearly on top of each other; our beds practically touched at the frames, and my poor parents had no privacy. My mother was never one to complain, but here she complained. My father, too. And yet Philippe and I could not have cared less. If it had been up to us, our quarters could have been tighter still, for we were all together in a time of war and upheaval. For us, this was all that mattered.

3

LIQUID AIR

Throughout the war years, my father maintained his great friendships with Philippe Tailliez and Frédéric Dumas. This alone was no surprise. Dad always said it would take more than the Germans to break up les mousquemers. Indeed, from my perspective as a small boy, the three men seemed as close as brothers. In my memory, they were a constant presence in each other's lives—and, consequently, in mine. However, the nature of their relationship changed during these consequential times. Because of their expertise in and around the water, they were in a fairly unique position. They continued to swim and dive, and my father continued to make primitive attempts to record their exploits on film and to improve upon their experiences in the water—but now they were able to pursue their passions in support of the French Navy. Now there was a sense of mission to accompany their passion and purpose.

Not long after my father left his active-duty post and began working in the secret service, the navy established a research division known as *Groupe d'Etudes et de Recherches Sous-Marines*—or, GERS for short—with Philippe Tailliez in charge of the operation. Judging from the stories I heard from my father and his friends over the years, it was a scattershot enterprise. Oh, they did essential work, the men were quick to point out, but they did so with a sense of élan and devilishness one did not often encounter in a military setting. I suppose it would be fair to suggest that they were cut from a different cloth than other servicemen.

There was one famous story that was a particular favorite of mine for the way it showed my father and his fellows as they truly were—young men at war and at play all at once. For some reason, there was an old upright piano aboard one of their ships, and for some other reason, the men used it for target practice. How this practice came about was never made clear to my young ears, but it was presented as a natural progression: If you place a piano on the deck of a ship, then it surely follows that you and your fellow officers are meant to shoot at it. And so these GERS officers would fire at the weathered instrument, which would offer a resounding, discordant hiss in response, and the men would laugh and shout like savages with each shot. Surely, the juxtaposition of war and music was not lost on them.

Didi remained a civilian, so he was not directly involved in the group's efforts, but as the undisputed best goggle diver in France, he was frequently nearby to offer an assist. In fact, Didi was so "indirectly" involved that he appears in many of the photos that survive from this period, alongside my father and the other GERS divers. Even though he wasn't active in any sort of official capacity, he most assuredly played a pivotal role—yet another example of my father's great friend, turning up like a lucky penny, always present and accounted for.

What this meant for my father, beyond the counterintelligence gathering that could now take place beneath the surface of the water as well as above, was that he was encouraged in yet another of his extracurricular pursuits—namely, the development of a new technology to assist in underwater breathing and to bring him closer to his long-held dream of being able to swim unencumbered among the fish of the Mediterranean. He was not the first to think along these lines; he was not even the first Frenchman, we would later learn. Like many "amateur" divers splashing about in the Mediterranean and elsewhere, he was not satisfied with the developments in this area. Now that he had engineered a workable housing for his underwater photography and had become fairly expert at dismantling his cameras each evening to clean and dry the salt water from the various moving parts that would invariably get wet, JYC was becoming increasingly frustrated with his inability to keep up with Dumas. Didi's exploits underwater had quickly become legendary: His skills as a spear fisherman were unsurpassed, and his superhuman ability in the water so captured my father's envy and imagination. My goodness, the man could skin-dive to depths of more than 65 feet, and his lung capacity had only increased with the years. Unlike other "amateur" divers, my father felt a particular disappointment, because his constant diving companion was so much more capable in the water.

My father worked with the GERS toward the end of the war, but throughout the early 1940s, he found time to work on his diving theories, to perfect his underwater filming capabilities, and to generally pursue his dreams. It was an odd mix of personal freedom and military constraint. Here, again, I recall very little of these moments in any kind of firsthand way, but as I grew older and began to pay more attention to my father's singular accomplishments, I began to recognize this time in his life as pivotal. He himself spoke about this period quite often, as I developed and expressed an interest in his work. Indeed, most of his initiatives during this early phase took place on the government's time, but he also filled his own time with experiments and studies and calculations. It seemed to me and my brother that Dad was a flurry of perpetual motion, always moving toward some objective or another, never quite satisfied with where he was, and off in search of something new, something better.

More and more, my father grew desperate to find a way to remain underwater for longer periods of time—in part to keep pace with Didi, but more than that to fully experience the wonders of the deep. In this, he was driven. He was not particularly fond of the so-called "hard-hat" diving technology, the state of the art at the time, which permitted determined undersea explorers to roam the seabed encased in a cumbersome metal helmet, fed by air tubes running to a surface ship directly above. The device, perfected by the legendary naval captain Yves Le Prieur, was described quite famously by Jules Verne in his classic book *Twenty Thousand Leagues Under the Sea.* According to my father, the device was more suited to science fiction than to the realities of underwater exploration. You could not swim, of course, when fixed to one of these contraptions. You could not move about horizontally, like a fish. You could not explore a shipwreck or swim through to the other side. You could not enter a cave or an underwater garden of rock or corral. You could only advance with tentative steps and keep tethered to the ship at all times. To a man like my father, who longed to experience the thrill and freedom of swimming among the oceans' creatures, the technology was more limiting than liberating.

At first, my father thought there might be a way to feed air to a diver through a hose, without having to do so from a stationary tank on the surface. It was a simple solution to a complex problem. Ideally, JYC believed, divers should be able to carry a tank with them as they swam underwater; however, the use of air in this manner, subject to the additional pressure, would be impractical—and, most probably, dangerous as well.

During one of his wartime visits to Paris, Dad mentioned his frustration to his father-in-law, and my grandfather put him in touch with a man named Emile Gagnan, an engineer at Aire Liquide. This very association would lead critics to later remark that my father had engineered his relationship with my mother for the sole or even sidelong purpose of accessing Henri Melchior's contacts in this area, but the timing indicates otherwise. And so does my parents' relationship. To my mind, my parents most certainly married for love—a child can tell, in his own way. Over time, my father would demonstrate certain weaknesses as a partner, as I will relate a bit later on in these pages, and I'm sure my mother could be a difficult companion on occasion, but this introduction to a key engineer was merely an outgrowth of the relationship. And, it was good business, because my grandfather thought the two men could be helpful to each other as well as to the fortunes of Aire Liquide. In this, he was surely prescient.

At the time, Gagnan was working on a regulator designed to help firefighters who were often unable to breathe efficiently while they were fighting fires due to a lack of oxygen. (The flames, as many of us remember from our high school science classes, quickly swallow the oxygen, especially in the confined spaces of a burning room.) It turned out to be a fortuitous introduction, because the two men hit it off and began to compare notes. Dad found a way to express his passion to Gagnan—no surprise, really, because my father's enthusiasm for an idea or project could often be infectious. Soon, the two fast friends came up with a way to release compressed air underwater, through a demand valve connected to a chamber. The device, soon and forever known as the Cousteau-Gagnan regulator, was similar to the valve Gagnan had designed for use above ground. The significant difference was that it could deliver air at ambient pressure—an essential development, because it meant that divers could now breathe underwater with the same ease and comfort as they could on dry land, at atmospheric pressure. And the demand aspect of the valve was central, because it would happen automatically. Together with the regulator, the two men designed an accompanying harness and mouthpiece, which was connected by double hoses to a single tank the divers would wear on their backs, allowing them to swim unencumbered. Divers had only to keep the mouthpiece in place and continue to breathe regularly as they explored the oceans' depths—untethered, unencumbered, unbound. It was altogether an inspired piece of engineering, and a significant advance beyond any existing technology. The two men were quite pleased; they could see the practical and transformative applications almost immediately.

They tested the device in the murky waters of the Marne River—an ironic choice, I have always thought. Why test a mechanism designed for use in the open sea in fresh water on the outskirts of Paris? Further, the results of these first test "dives" must have surely disappointed, for my father had always dreamed of swimming freely in a blue or turquoise sea, overwhelmed by the magical colors and vibrant ocean life all around. The very first time he opened his eyes underwater, as a boy in the muddy waters of Harvey's Lake in Vermont, he could not see for more than two feet; in the clear shallows of the Mediterranean, he could see for vast, breathtaking vistas. Here in the Marne, the water was so cloudy he could not have glimpsed the reaches of his own outstretched arms, and yet there was enough magic and hope and possibility tied to the simple act of breathing underwater that these absurdities did not seem to matter much to these two determined men.

My father knew that it was not sufficient to merely supply air to divers to allow them to breathe underwater. You see, human beings are accustomed to breathing at atmospheric pressure, but as you descend deeper into the water, the water itself exerts increasing pressure on the chest and lungs. For most people, it becomes difficult to breathe through a tube at depths of more than three or four feet. At certain depths, it becomes impossible to breathe at all, and this explains why Frédéric Dumas's ability to skin-dive to depths of approximately 65 feet was so impressive—and, so unusual. Indeed, the very first underwater film my father ever produced was titled *Par dix-huit mètres de fond*—or *Sixty Feet Down*, that is, 18 meters. The title referred to the generally accepted limits of human endurance—limits JYC and his cohorts now hoped to shatter. Most divers were not built in the same way as Didi, and so for this reason, it was necessary to use a regulator of the type that my father and Emile Gagnan imagined to match the surrounding or ambient pressure, to allow divers to fully explore the vast depths of the world's oceans.

All of a sudden, Dad's desire to swim like a "manfish" was very nearly a reality. He wrote quite movingly about his first diving experiments with the regulator in *The Silent World*. The initial dives with the perfected apparatus took place in June 1943, in a sheltered cove not far from Villa Barry, as far away as possible from curious swimmers and the Italian troops occupying the area. My father looked forward to these experiments for months, as engineers at Aire Liquide constructed a prototype based on the designs he made

with Emile Gagnan. Years later, when I was old enough to understand the significance of the moment, JYC told me that the wait for
the first prototype apparatus was fairly interminable. However, he
also reminded me that good things do indeed come to those who
wait, and here he was like a child at Christmas, counting the days
until he could open his presents. I do not remember that interminable
ble wait in any sort of firsthand way, but I do recall the day it ended:
A wooden case arrived from Paris, containing the final Cousteau-
Gagnan prototype, and the three mousquemers tore into it with
abandon. There was such shouting and merrymaking and jubilation
that it felt to us kids like the men had won some fabulous prize.

Today's divers commonly use a single-hose mechanism, attached
to a single pressurized tank. However, the Cousteau-Gagnan regulator worked with a twin-hose design—or "aqualung"—and les
mousquemers tried it on straightaway. They took turns marveling
at the moment. My father could hardly contain his excitement; the
others were energized as well. It was to be another grand adventure, played out in another grand setting, to another set of grand
results. I don't think my father or his friends cared a whit about
the moneymaking opportunities their new device might present;
they cared merely for the freedom that lay in wait. The plan was
for JYC to make the first trial dive himself, with Tailliez, Dumas,
and my mother stationed at strategic points to offer support. It was
my mother's job, for example, to swim out some 300 feet or so and
observe my father's progress from the surface, with a diving mask
and breathing tube. She would look down into the water and witness my father's "maiden voyage," well positioned to signal Didi if
Dad appeared to be in distress, while Tailliez would monitor the
proceedings from the shore.

"I looked into the sea with the same sense of trespass that I have
felt on every dive," my father wrote of his first "regulated" dive in

one of the opening passages of *The Silent World,* describing the scene: "A modest canyon opened below, full of dark green weeds, black sea urchins, and small flowerlike white algae. Fingerlings browsed in the scene. The sand sloped down to a clear blue infinity. The sun struck so brightly I had to squint. . . . I looked up and saw the surface shining like a defective mirror. In the center of the looking glass was the trim silhouette of Simone, reduced to a doll. I waved. The doll waved back at me."

I was a boy of 15 and already an experienced diver myself when I first read those passages, and the imagery made quite an impression. Now deep into adulthood, I still think of my father's words every time I look to the surface at the end of an exhilarating dive: ". . . shining like a defective mirror."

It is an apt metaphor, wouldn't you agree?

As I have indicated, my father was not the first diver determined to swim like a fish. Prior to the collaboration of Jacques Cousteau and Emile Gagnan, there were at least two documented examples of other pioneers in this area—each engineered by a Frenchman and tested in a river in the southwest and in a swimming pool in Paris and each manufactured for commercial use with only disappointing results. The first, a simple tank and mask apparatus, had its origins in southwest France in the 1860s; it was developed initially for coal miners, to offer an additional supply of air in the event of a mine collapse. Later it was adapted by a French naval officer who thought it might prove useful for placing mines beneath enemy ships in a time of war. We took pictures of the device for a documentary I produced entitled *The Manfish,* to honor the phrase my father and his friends attempted to coin after their first successful experiments with the Cousteau-Gagnan regulator. It really was quite a clever contraption. Crude, but clever. I tried one myself, on that same southwest river where it first had been tested, and found that it was somewhat effective even after more than a hundred years.

Then, in the early 1930s, not long before my father took up free diving himself, another French naval officer essentially had the same idea, only his contraption was rigged in such a way that the tank was fixed to the diver's chest. It was awkward and burdensome and not especially conducive to the free, unfettered motion of swimming. Also, the diver had to manually control the release of the air, which was very inconvenient and required the diver to regulate his own breathing. Here, again, the device was somewhat effective and certainly an improvement over simple human lung capacity, but it was nothing like the innovation of the Cousteau-Gagnan demand regulator, which permitted divers to swim for extended periods, without constraint.

Another irony: These innovations were each perfected and refined in fresh water, out of reach of the salt of the sea, just as the Cousteau-Gagnan regulator was put through its initial paces well inland. Now, though, Dad was ready to put his device to the real test. Following that first test dive in the cove off Toulon, it was impossible for les mousquemers to contain their excitement. Such a triumphant moment. Such possibility. The world as they knew it would be forever changed—the undersea world most especially—and along with it, there would be a new terminology. In addition to the term *aqualung,* which soon fell into common usage, the complete set of equipment came to be known as a self-contained underwater breathing apparatus—or, by the acronym SCUBA. Soon, the acronym itself seeped into everyday conversation.

In just a short time, the word *scuba* would come to mean the same thing in any language, although to my father's ears, it meant only one thing: freedom. For the rest of his life, he always said he would hear the term and it would remind him of these first experiments. He would hear it, and it would be like music, he explained, because it was so inspiring.

Of course, there were improvements to be made in the initial design, and there needed to be further study regarding the limits of the human body in conditions that barely resembled the atmospheric pressure to which humans were accustomed. Already, my father and his friends and the loosely assembled diving community to which they suddenly belonged had developed a series of safety guidelines and precautions. They did this by applying their knowledge of physiology to their own firsthand experiences, which they had collected by trial and error. They consulted with naval doctors and compared notes with other pioneering divers. They were quite pleased to learn of a scientist named A. R. Behnke, whose fine work in diving decompression models offered a template for their own research. Soon, they understood the different stages of *l'ivresse des grandes profondeurs*—the seizures or "intoxications of the great depths."

They even experimented with various ways to keep the human body warm during long hours spent in frigid waters. They had never had a problem diving in the Mediterranean, where the air temperatures are mild year-round, but the waters at the lower depths of the sea are particularly cold, and this was proving especially difficult and worrisome as they spent more and more time underwater. At the time, conventional wisdom suggested that coating the skin with grease was an effective way to combat heat loss, but les mousquemers had quite a different experience. They found that when they dove at great depths for extended periods they would actually lose body heat more rapidly when they had applied a thick coating of grease than when they had not. Soon, they fashioned a rubberized suit, which was only somewhat effective in this regard. It kept them warm, but it was by no means a formfitting design, so great pockets of air would invariably bubble beneath the rubber and make it difficult for the diver to swim evenly. (To counter this, they tried swimming with weights, which also proved beneficial when they were skin-diving.)

The first and foremost rule of diving, they determined, was to never dive alone. "A person alone is in bad company," my father used to say—and this was surely so. Happily, among us Cousteaus, this was never the case in those heady early days that followed my father's initial experiments. We dove together, constantly. I was only six or seven years old, but already I was an experienced swimmer. I caught on soon enough. Philippe, too, was swimming like a "boyfish" before long. We were as comfortable in the water as we were on dry land, and I believe it was a source of great pride to my father that his children were so capable in this regard.

I watched joyfully as my father and his friends became more accomplished in the water and as they perfected their design. My mother was also soon an expert—certainly, one of the first female scuba divers of any note or accomplishment and the very first to even make an attempt. This last was a distinction I never quite understood, because she followed the first male scuba divers by only an afternoon. She was not just "the doll" waving to my father from the surface on his first trial dive with the aqualung; she donned the equipment herself and took to the water like a mermaid. Philippe and I often watched their happy escapades from the harbor shore—or, just as often, we'd accompany them on their small boats as they positioned themselves out at sea for deeper, more adventurous dives.

The Toulon Harbor and the bay of Sanary were my father's playgrounds—and ours as well. And the liberating new technology was not just for play. My father took the opportunity of these welcome developments to improve his underwater filmmaking techniques, putting them to good and productive use to record a variety of local "expeditions"—including one effort by an Italian salvage diver to survey the wreckage of an oceangoing tug that had been downed with the rest of the French fleet, which my father had taken to calling "the suicides of Toulon." The *Dalton* now lay in 45 feet of clear water in the outer harbor and had become overgrown with weeds and algae in less than a year.

From there, he expanded his underwater subjects to include other legendary wrecks, such as the battleship *Iena,* which had sunk during the First World War, and the freighter *Tozeur,* which was sunk by a treacherous wave off the coast near Marseille during the war that had just ended.

Dad's filmmaking continued with a renewed sense of purpose— but so did his diving. Even as he filmed hard-hat divers on salvage missions, deployed by the navy to examine these downed vessels, he was swimming freely with his companions, enjoying their new expertise and their burgeoning reputations as the first and foremost menfish of the region. Soon, we were diving as a family. By 1945, we each had our own set of equipment, which had by now been greatly improved. I was barely seven; Philippe was four and a half; we were both strong swimmers. How could it have been otherwise? We had grown up by the sea. There is an out-of-focus photograph among my possessions of my father "throwing" me into the water, taken around this time. Like most children, I was in the habit of approaching the water tentatively; I could swim, quite well in fact, but I took my time easing into the water. Just before this picture was taken, my father came upon me from behind and made quick work of my approach. We look rather alike in this particular photo—me, dressed for a dive like JYC in miniature; my father, smiling mischievously as he "tossed me overboard." From time to time, I find myself gazing at this picture, lost in thought, imagining the young man my father was, the joy and abandon he and his friends felt at their daily discoveries, and the hope for the future he must have seen in me and my brother, Philippe.

The transformation was very nearly complete. My father had reimagined himself as a manfish, and we quite naturally went along for the ride—soon enough, whenever and wherever we could. For us, diving and exploring below the surface became a natural extension of our time on land. It was how we were and who we were as a family. Every weekend, it seemed, we were out on the water. Holidays, vacations—it was much the same. Most of our early diving experiences were in the south of France, somewhere off the coast between Toulon and Marseille. Occasionally, we would head out to any of several islands just off shore. My parents were always looking for caves and coves and shipwrecks and unusual underwater habitats that we could explore. The idea was to seek out something new, something no one had ever seen before. At the same time, we kept floating back to more familiar habitats in the cove just outside our own front door.

Indeed, the rocky beach across from our modest home at Villa Barry became such a focal point of our first dives that the French government eventually placed a plaque there, to mark the spot where Philippe Tailliez, Frédéric Dumas, and Jacques-Yves Cousteau conducted their very first diving experiments—to honor the three musketeers of the sea at their point of entry. But we did not need a plaque to announce the sea change that was taking place in our own backyard. Everyone who lived and worked and visited that harbor in those days can attest to the swirl of excitement that accompanied les mousquemers as they bounded toward the sea. I can close my eyes and still picture them: three young men, brimming with joy, exhilarated by their underwater adventures and their time together, spilling into our "backyard" slice of the Mediterranean like it belonged to no one else. And yet, what a lot of people don't realize is that it wasn't just the mousquemers reveling in these new underwater discoveries; it was a family affair that reached to the rest of us as well and extended ever outward to include a happy band of fellow divers that soon numbered in the dozens.

I continued to dive with those three men for the rest of my life—until they disappeared one by one from this life and on into the next one. As ever, I do not wish to get ahead of my story, but I believe it's appropriate to reflect here on that lifetime of diving alongside these good men, who in the silvery eye of my memory would forever be the three young men of the sea. I look at photographs from their GERS days—so young, so vital—and these are the freeze-frame images I carry with me. Ironically, it was Dumas, the youngest of the mousquemers, who was the first to leave us— possibly because he was a lifelong smoker. Ironically, tragically, poetically—his death struck Captains Cousteau and Tailliez with every conceivable emotion, just as the subsequent deaths of the latter would fill the hearts of those they left behind. Didi's famous lung capacity would prove no match for a lifetime of tar and nicotine, but he kept up with his diving until the very end. My father was next, in 1997, and he too continued to dive until he could no longer walk.

Tailliez, the oldest of the group, was the last to leave us, and we remained cherished diving partners until the very end. I held on to his friendship like a lifeline, for he was the last link to the young man my father had been when he pried open that wooden crate from Aire Liquide, to the discoveries we all shared—the wonders so many people now take for granted. Captain Tailliez held fast to our friendship as well—I suppose, for his version of the same reasons. In fact, he called me on the telephone just before his 93rd birthday to invite me to a special gathering. He said, "Jean-Michel, are you coming with your dive bag?"

The question took me by surprise. "What do you mean, Captain?" I asked. I had known Captain Tailliez since I was a small boy, so of course I referred to him with the honorific he so deserved.

"We're going diving," he said.

He might have been 93 years old, diving's eldest elder statesman by at least a few decades, so who was I to refuse such an invitation? I brought a cameraman with me to the celebration to record the event for posterity, and sure enough, we went for a fine dive in the same waters where the captain had swum with my father and Didi all those years ago.

The following year, I got a similar call, only this time Captain Tailliez was not permitted to dive, according to his doctors. It was not for merely chronological reasons, I should point out. And, happily, it was not the beginning of the end for my father's dear friend; it was merely a swell on the sea of a life at twilight. Philippe had suffered a severe cut a few days earlier, and his doctors were worried he'd develop an infection if it was exposed to the salt water, so we took him down in a submersible and filmed that instead. His spirits were high despite what we all hoped were his near-term limitations. We sang "Happy Birthday" and made it a joyous occasion, vowing to resume our tradition the following year.

Sure enough, the call came again the next year. This time, Captain Tailliez was turning 95. "Are you bringing your dive bag, Jean-Michel?" he said, in the familiar sea-soaked voice I had come to know so well.

"Yes, Captain," I said. "And my camera."

"Good," he said. "We are going to dive where I first dove with your father and Frédéric Dumas. It will be our tradition."

So I traveled once again to Bandol and met Captain Tailliez with my crew, only this time I was struck by how much had changed since I was a boy—since my last visit, even. It wasn't Tailliez who looked so different to me; it was the sea itself. We splashed about in the water, but there was almost nothing to see. Everything was dead, or dying. The entire harbor had been fished out. It was such a sad situation—a disaster—but I could almost look past it in consideration of my steady diving companion. That we were there at all, diving in these same waters, was a cause for celebration—a celebration amid such tragic devastation, to be sure, but a celebration just the same.

At one point, our heads bobbed to the surface at about the same time, and I swam over to my friend and collected him in an affectionate hug. "Captain," I said, "my dream would be for you and me to return to these waters every year on your birthday. I'd like to come back to this very same spot and do the same dive when you turn one hundred."

At this, Philippe Tailliez looked at me and said, "But Jean-Michel, you are already an old man yourself. You might not be around!" Then he laughed heartily, and it was as if the years that had passed since we'd first arrived at this place had suddenly dipped beneath the surface of the sea.

Such a rascal.

Sadly, Philippe Tailliez didn't make it to one hundred years. He died two years later, at 97, but I offer this exchange here for the way it solidifies the place this man held in my father's life—and, by extension, in mine.

Reaching back to those early dives, when my family was embarking on our lifelong adventure in the water, I am transported. I am a child once more, swimming for the first time with my father and my new underwater gear. He threw me overboard into a sea of infinite wonder. So many years have passed, and yet the memories are close enough for me to touch, as if they happened yesterday. I remember Dad teaching me to fit the mouthpiece behind my lips and reminding me to try not to talk. Already, my brother, Philippe, and I were experienced mask divers, but now the first thing we wanted to do was express our amazement to each other while we were underwater. It was a classic beginner's mistake—and here we were, the very first "beginners" in the history of deepwater diving. We kept trying to talk to each other, to shout with excitement at what we were seeing. Each time we did, we would get a mouthful of water and have to scramble to the surface to catch our breath and refashion our masks.

In contrast to the stark, later dives with Captain Tailliez in that same cove, in those days, we saw a lot of fish—lobsters and octopus, most memorably. I remember playing with the octopus, in a manner taught to us by Dumas. We would grab them, and they would spit their ink. They would try to swim away, and we would catch them. We would watch Didi fool for long stretches with an octopus, and then that one would grow tired of the game and drift away, only to be replaced by another one soon after. Of course, we were just impressionable children, so we tried to mimic our great friend Didi at the first opportunity—with happy success, it turned out.

One of the great contradictions of my father's often lyrical descriptions of his adventures beneath the surface was the way he called his first book *The Silent World*. He placed the same title on one of his first documentaries—a different story, although he gave it the same name. But there was nothing silent about it. Yes, it took some getting used to, swimming off the ocean floor. There is quite a lot of *crackling* and *whooshing* and jostling about. A lot of people don't realize this until they experience it for themselves. And yet, compared to the noise and clamor of the aboveground world, I suppose there is a kind of silence—a clamoring silence, but a most definite silence. A silence I came to treasure as it became more and more familiar.

My father often spoke of this so-called silence. It was a silence he longed to hear.

4

A FAMILY AT SEA

O ur world on land was not nearly so quiet—or, so predictable. We had our share of noise and clamor. We had
our unwelcome surprises, too. I look back on those war years
and think first of our fine and thrilling experiences in the
water, diving and exploring, but there is also our day-to-day
experience to consider. The war had been disruptive to our
life as a family, just as it had unsettled families throughout
Europe. Our own makeshift dynamic was shaken as well, as
my young cousins Jean-Pierre and Françoise were pulled from
our life almost as swiftly as they had joined it. This coincided
with the end of the war, and my family's almost total immer-
sion in the water and served to remind us that there would
always be weighty, earthly matters to consider above ground,
even as we swam weightlessly at the bottom of the sea.

The tricky business of keeping one foot (or fin) in each world, above the water and below, would mark our lives going forward.

Some further background is necessary. My father's brother, Pierre-Antoine—my cousins' father and my uncle—was a journalist, on the wrong side of the war. My uncle was known to friends and family as PAC, blending his initials in much the same way that Dad took on the nickname JYC—in no small part to emulate his older brother, I always felt sure. (Indeed, in adolescence, my father often called himself Jack, which sounded very conveniently just like PAC, and so reminded any and all of their special relationship.) PAC was quite a different character from Jacques-Yves, even though they had both grown up in the same household, under the same influences. In many ways, their differences were on full display in the routines of their children and the choices they made on our behalf. When we all lived together in that small house in the Alps, for example, my grandmother would offer lessons to Jean-Pierre and Françoise, because it was my uncle's wish that his children continue with their schooling throughout their displacement. And yet Philippe and I endured no such lessons, because our parents did not wish for us to receive a formal education at such a young age. They did not see the need. In fact, they put off enrolling us in any kind of school setting for as long as possible. There was plenty for us to do and learn and experience, they believed, without the formality of books and lessons, so while Jean-Pierre and Françoise endured their homeschooling sessions, Philippe and I were free to romp and play and make mischief.

These contrasting points of view made for an interesting tug and pull within our small, thrown-together household. Already, there must have been a good deal of tension between my mother and her mother-in-law—my grandmother, whom we all called Mutti. I did not recognize this at the time, but I have thought about it often over the years and have come to the conclusion that to a young woman in her 20s, the circumstances must have been impossible; to a confident young grandmother, accustomed to having her say, untenable. I even discussed it with my mother—years later, as adults—and I had to admire the way she steadfastly held her ground against such a forceful and resourceful presence as my grandmother. It could not have been an easy situation, for either woman. In many ways, my mother was still so young, and here she was consigned to a solitary life in the mountains with her husband's mother and his brother's children. It must have pained Mutti as well to see my mother struggle with two small children and a mostly absent husband in much the same way that she had struggled with two small children and a mostly absent husband. They were a lot alike, but here they were at different points on their line of experience.

In this setting, in this context, I always thought of my grandmother as the adult in the relationship. She was serious and strict, whereas my mother was more carefree and unpredictable. When Philippe and I fell into fighting, as rambunctious boys invariably do, my mother's style was to let us wrestle until we grew tired of it and to settle our differences in the manner of men, whereas Mutti would look on and *tsk, tsk, tsk* her disapproval. And she did not have to *tsk* when it came to disciplining my cousins, for she was not only their grandmother, there to lend a helping hand in a time of need, but their de facto guardian as well. She was grandmother to one set of children and mother to another—all at once, all under one roof.

The two women divided up their household responsibilities along fairly even lines. Despite her many fine attributes, my mother was a terrible cook, so Mutti held sway in the kitchen. My mother kept up with the shopping and the housekeeping. And both would see to our discipline, in their own manner. Our circumstances were certainly unlike any my mother could have ever imagined when she married my father, a man of the sea who promised a lifetime of adventure. This wasn't turning out to be the case just yet. Young Simone Melchior Cousteau hadn't expected to be living with her husband's mother and two children not her own, together with her own two children in a landlocked home in the mountains while her husband was stationed elsewhere. But she accepted her role and her circumstance with great good cheer; she was willing to do what needed to be done.

Meanwhile, my uncle was forced to flee France as a result of his misplaced sympathies for General Pétain, who had openly supported Hitler's campaign, through the offices of his newspaper, *Je suis partout* (I am everywhere). He was believed to have found sanctuary in Germany, but we had little communication with him once he left the country, and so Mutti now became his children's legal guardian as well. It meant there were now two different voices of authority in our busy household—my mother's, in relation to me and my brother, and Mutti's, in relation to my cousins. And so the tug and pull continued. Soon after, we all moved from the house in the Alps to our more familiar territory near Toulon, in Bandol. My grandmother decided to return to Paris with Jean-Pierre and Françoise, where she believed they could receive a better, fuller education and where she could expose them to all manner of cultural diversions—but our time together had indelibly stamped my growing up, just as it had been an important marker for my cousins.

To me and my brother, it was as if we were losing pieces of our childhood. Jean-Pierre and Françoise were like our siblings. We did not have a lot of children in our midst, so they were our constant playmates. We had been together for as long as any of us could remember. We had squabbled and conspired and teased. We did everything together—beyond our differences in schooling. We even swam together and took our first rudimentary dives in one another's company, before Philippe and I began to outpace our cousins in this one activity. Together, we had been afraid when a group of armed Italian soldiers spilled from their trucks by the gate in front of our house in the south of France, disrupting a guessing game we used to play whenever we heard a passing vehicle. *Who is driving?* we would take turns wondering. *Where are they going? How many people are in the car?* One of the soldiers thought to amuse himself and his fellows by pointing his gun at us, and we cousins were fairly terrified, but then the soldiers moved on, and we were left to talk about the incident for weeks and weeks. The way we remembered it became another one of our games, just as we had often amused one another with games of imagination. Mostly, we had imagined our lives after the war, when Jean-Pierre and Françoise could be reunited with their parents, and we could be a big, sprawling family—together, still, beneath an even bigger roof, sharing bigger adventures.

But Mutti had a different idea, and we could only respect her decision and make our good-byes. It was a particular sadness for us cousins—but, again, it was a time of war, and there was particular sadness all around. We simply accepted it and moved on.

The move left me and my brother to each other. We were a little more than two years apart and fairly opposite in terms of personality. I was a me-first sort of child, always grabbing at the world and its many riches as if I had good fortune coming. With this in mind, my exasperated mother took to calling me Jean-Mine, for the way it reflected my self-centered worldview. Philippe was more circumspect, more collaborative in nature, more inclined to accept good fortune than to reach for it. Even our physical appearances were in contrast. As a boy, I was tall and slender, almost thin, with an aquiline profile that featured the famous Cousteau nose. Philippe was more heavyset, and in this way, far more Melchior than Cousteau. His nose, just to cite one particular feature, was rather broad and flat. I'm afraid it cut such a different picture from me and our cousins that

I often joined Jean-Pierre and Françoise in teasing my poor brother about his bulbous nose.

Without Jean-Pierre, with whom I had been especially close because of our proximity in age, Philippe and I were now drawn to each other. And, looking back, I believe our new bond was cemented by our time in the sea, discovering the vast wonders of the deep with our parents and thrilling to the many adventures we would now experience as a family. This last was key, because now that Europe was at peace, my father's time was once again his own, and this marked a truly wonderful period in our lives. More and more, Dad was a real presence in our household. More and more, our family moved to his rhythms. In the mountains, he had been back and forth for visits whenever his schedule allowed; even here, back now in the south of France, his time with us had at first been spotty, but as the war finally and mercifully came to end, so, too, did his unpredictable schedule. He was a fixture in our lives once more—or, I should say, at long last, because I could not remember a time when he lived with us on anything resembling a permanent basis.

We soon settled in a new house in Sanary-sur-Mer—a little farther to the east, closer to Toulon. The house was known as Villa Reine, and it was one of several houses in the area that had been taken over by the French government after the war and made available to government employees and military personnel—and, in this case, naval officers. It was a big house in the middle of a forest of tall pine trees not far from the shore. Unfortunately, we could not see the water from the house, which was a marked change. Already, we had become so accustomed to the sea that my brother and I expected to look out on it from our bedroom window. It was a dispiriting shift in outlook, because it meant we would actually have to make an excursion to get to the sea, instead of spilling out our front door and bounding those few steps to the shore.

Still, we were close enough—however, here at Villa Reine, I would occasionally notice that my father and his friends were as adept at cavorting around in our backyard as they were in the sea. They were such fine, strong athletes that no environment could contain their exuberance; even my father, who had been maladroit as a young man, was now graceful and muscular. Once, I looked out my window to discover my father and his friends attempting to be high-wire artists in our backyard garden. It was an astonishing thing to see—grown men, traipsing across a tightrope they had strung between two pine trees. Didi was among them, of course; he was ever-present, a fixture in my father's life and now just as surely in ours.

My father explained later that these afternoons of acrobatics offered a good way for him and his diving companions to keep conditioned and to build their agility for their time underwater. Even as a child, I recognized that it was mostly a way for the men to roughhouse and have fun and amuse themselves in a manner of extremes.

There was nearly a calamitous side note to this memory. Prior to attempting their acrobatic display, my father and his friends had placed a mound of pine needles beneath their tightrope to serve as a cushion in the likely event of a fall. The rope itself was not strung particularly high off the ground—about three or four feet, as I recall—but the ground was hard, so a cushion was a sound precaution. However, my brother had just discovered the impish properties of a book of matches, and for an unfortunate stretch of several days was fond of lighting matches and tossing them to the ground. On most days, nothing much came of Philippe's particular mischief; as often as not, the wind would blow out the matches as soon as they were struck. But one day, a series of events unfolded that gave us all—me, most of all—a serious scare.

I happened to be allergic to caterpillars, and the gathered pine needles were like a magnet for these pests. If one of them touched my skin, I developed a rash that would cause me to itch to the point of bleeding. It was nothing serious—but a serious nuisance, just the same. Somehow, we'd discovered that sprinkling a bit of gasoline on the caterpillars would turn them away, in search of other places to nest. I could not always do such a neat job of it, however, and sometimes sprinkled a bit of gasoline on myself—a particular hazard when your younger brother is in the bad habit of lighting matches.

As my preamble to this story suggests, one of Philippe's matches caught a brush of dried pine needles in precisely the wrong way as I crossed nearby at precisely the wrong time. In an instant, I was lit up like a human torch. Now, all this time later, I can look back and laugh at our stupidity and dumb luck, but it was certainly a precarious moment. Fortunately, my mother happened to be in the yard hanging the laundry, and she quickly threw a wet blanket on me and smothered the flames before there was any lasting damage to anything but my clothes and my hair—and, I suppose, my pride.

Of course, Philippe and I were soundly and justly punished. It might have been Philippe who was primarily at fault, but as brothers, we gladly took what the other had coming. We were co-conspirators—and, therefore, co-defendants. My mother told my father what had happened when he came home from the sea that evening, and then he turned to us and said, "Tomorrow, at six o'clock." It was a simple pronouncement, but we knew full well what it meant. It was his way.

To reiterate, my parents often meted out punishments to us kids. In fairness, that was how most children were raised at the time—in France, as elsewhere. However, they would never spank us in anger, which I believe was the reason my father always announced one of his "appointments" for a deserved beating. That was the routine, and I now see the wisdom in his approach. My father was a man of great passion and character, but I would not list anger as one of his character traits. Even on one of his expeditions, where much was often at stake, he would not lose his temper when the missteps or miscalculations of one of his crew might have cost him dearly, in one way or another. He would step back from the moment and gain some perspective, which was what he appeared to be doing here. He set the time ("Tomorrow, at six o'clock"), and we were all meant to fill it with reason—Dad, to think through what he was about to do; me and Philippe, to think about what we had just done.

Mostly, though, Philippe and I would look ahead—to our next adventure. Our time in the water was becoming everything to us, not least because it gave us such precious moments with JYC. We were like little fish. My mother and father were like big fish. Together, we swam about in the only "school" my parents deemed necessary, and soon our transformation was complete. We would be as comfortable in the ocean's depths as we would be on dry land—only more so, if such a thing was possible.

Very quickly, diving became a part of us, as if it had been hardwired into our DNA all along. It got to where, in adulthood, after a virtual lifetime at sea, I would not feel complete or whole if I had been away from the water for too long. There is no real winter in the south of France, so we dove year-round. As I wrote earlier, the water temperatures could be chilling at great depths, but they were mild enough, especially for short periods, close to the surface. We dove off the shore, off the pier, off a small wooden boat my father kept in a cove not far from our house. You could fit six or eight people on that boat, so we often went out with ten or twelve, such was the enthusiasm my parents and their joyful circle of friends soon developed for our dives.

Dumas and Tailliez were with us much of the time—not always, but frequently. Every weekend, it seemed, and very often in between. Unlike my father, these two were obsessed in the beginning with catching the fish we were now privileged to join in the deep. Didi would fish mostly for sport; Captain Tailliez, for money;

and JYC would keep them company in the water, either to record their exploits with his camera or to pass another few precious hours among his precious sea creatures. This would change soon enough, as our focus shifted toward conservation and preservation, but in the middle 1940s, there was such plenty in the sea that they hardly gave these matters a thought.

My obsession was more in line with my father's—that is, it was mostly a fascination. I have often wondered what it was with our diving that so captivated me and my brother and left us longing each morning to don our gear and pitch ourselves into the Mediterranean. How is it that two small boys took so swiftly and surely to the sea? The biggest attraction, I have come to believe, was the sense of freedom we could not help but feel as we swam among the sea's creatures and along the colorful seabed. It tapped into a basic human emotion—to soar, to defy gravity, to reach ever upward toward the heavens. Years later, I came across a famous sonnet, "High Flight," by the Anglo-American aviator-poet John Gillespie Magee, Jr., that touched on some of these same emotions: "Oh, I have slipped the surly bonds of earth, and danced the skies on laughter-silvered wings. . . ."

It was written about the time Philippe and I were learning to dive, but I did not discover it until several years later. Today, it's become quite famous. It's a line you'll find on a great many headstones in Arlington National Cemetery, and President Ronald Reagan even repeated it in a stirring speech he offered in the wake of the Challenger shuttle disaster. And yet, the first time I heard those words I felt as if they were my sentiments, describing my own state of mind as I swam in the sea, my own fascination and wonder, so I offer them here in an attempt to understand the siren pull we must have surely felt as we pursued our thrilling new activity.

Think of it: We humans have long been obsessed with flight. We are always looking heavenward, imagining ourselves as birds, soaring freely against a magnificent blue sky. As children, we are always looking up. We look up to our parents. We look up to the stars. We look up to receive food or attention. We never look down. We never look at the ocean. We never look at the ground at our feet. That's probably one of the abiding explanations for why we have never taken proper care of the planet. It's against our very nature. Sadly, as a result, we never look at the ocean. We are driven in such a way that we know more about the moon or the stars than we do about the world's oceans. Why? Because we are conditioned to look up instead of down.

My father was out to change all of that. At first, he meant merely to slake his own thirst for adventure and discovery, to experience the miracles of the sea for himself in a firsthand way. He wanted to record some of these experiences with his camera—and then, to perfect the manner of his underwater photography. It was only later, as he recognized the vast resource of our underwater world that he sought to popularize the exploration of the sea. It became his mission. He wanted others to know and experience what he and his fellow mousquemers had come to know and experience. He wanted us to look down—to make ourselves like fish and consider the world anew.

The contrast between my father's early, boisterous enthusiasm for diving—and that of his fellows—and the almost evangelical zeal he would later bring to his lifelong role as a kind of ambassador to our living seas was in evidence in their approach to the sport of fishing. In the beginning, it was very much a sport and a part of our activities in the water. We fished for fun, and we fished for food, and we did not spend much time wondering about any endangered species of fish or undernourished habitats. Conservation was not yet at the forefront of our thinking. That would come later. Frédéric Dumas and Philippe Tailliez in particular were ardent fishermen—and they would remain so for some time. In

Didi's case, I sometimes believed he fished for sustenance as well as sport, because he didn't appear to have any other noticeable source of income.

In this way, my brother Philippe and I fished as well, in the manner of boys eager to earn some pocket money. I fished mostly for octopus and shrimp. Soon, the enterprise became the focus of my daily activities. Even before I started formal schooling, I left the house early in the morning to tend to my catch. By the time I went to school, I would still turn up late on most days—very often in wet clothes that could not help but call attention to where I had been and what I had been doing. Fortunately, I had inherited my father's gifts of gab and flamboyance, and on most days, I was able to talk myself out of disciplinary action. I was usually let off with a stern warning to take my studies more seriously and to leave my maritime pursuits to after-school hours. As far as I ever knew, my parents were never called in to discuss my tardiness. They knew about my fishing exploits, of course, and the sideline business I had developed. It wasn't much, but it yielded enough for me to buy corks for my popgun, some sweets at the candy store, and an occasional treat at the local bakery. In fact, they encouraged me in this; they believed it showed a certain spirit of enterprise and self-reliance.

Somehow, I had developed a great respect for the sea and its creatures almost entirely on my own. Without ever being told to do so and with no regulations requiring it, I threw back every-thing I caught if it seemed too small. Perhaps I saw my father and his friends do the same thing, and I was merely mimicking their behavior. Perhaps I did this on the thinking that a paltry catch would yield an insignificant return when I attempted to sell it, but I like to think I did not wish to pull any fish from the sea before it had reached maturity. In this way, perhaps, I began to pay my respects to the ocean environment. In this way, I showed concern for the ecosystem before such a mindset even had a name.

I was helped along in my fishing business and my other seaside pursuits by my good friend Robert, who lived down from Villa Reine in the port of Sanary-sur-Mer. We did not have a telephone in those days, but Robert and I developed our own form of communication that proved quite effective. I'd climb atop one of the huge pine trees near our house and hold a small mirror to the sunlit sky. Then I'd tilt the mirror back and forth so that the sun's rays would bounce off of it in a kind of crude code. Robert could pick up the reflection from the balcony of his apartment on the harbor. We'd worked out two basic messages, as I recall. One: "You come to me." The other: "I'll come to you." There was nothing else to say, really. These worked well enough.

If we didn't arrange to meet at one house or another, our adventures would usually begin in the town square, in front of the church. That was our default meeting place. From there, we'd set off and fill our days with every distraction the seaside had to offer. We'd fish— some days for shrimp, some days for sea urchins, some days for octopus. (On one unforgettable afternoon, we ate so many sea urchins we were fairly drunk from all the iodine.) We'd swim. We'd sail. One of our favorite games was to see how many of us we could cram onto the small Charpi sailboat we kept in the harbor—perhaps, mimicking the boisterous enthusiasm of my father and his friends as they crowded onto our small wooden motorboat to go diving. Today, beginning sailors would equate the Charpi to an Opti—a small, boxlike shell with one sail, designed to hold one sailor. Oh, how we loved that little Charpi. Such a marvelous toy. And yet it simply would not do to put it to its intended use. We would not be solitary sailors. We would not be tied down by the constraints that bound everyone else. In this, I suppose, I was rather like my father, determined to go against the grain and cut my own path, so we'd gather our friends and squeeze onto the boat until we tipped over—one time squeezing as many as 11 of us onto that tiny shell.

From time to time, Robert and I, together with our friends, would get into some small trouble or other. We'd taken to calling ourselves the White Wolves and took great pride in knowing that the pretty girls in our age group often talked about our frivolous exploits in hushed tones. The biggest of these small troubles came one afternoon when the harbormaster hoisted the red flag to indicate that the waters were treacherous. Robert and I went out on the Charpi anyway, but then we capsized and had to be rescued by the patrol boat. The good news was that we were saved, but the bad news was that our sailboat had gotten away, so we had to double back to collect the vessel. To us kids, it was a terribly exciting misadventure, but to my parents, it was an example of my reckless, irresponsible behavior—the first of many, I'm afraid, although the extent of my recklessness and irresponsibility was certainly benign. In any case, I was in no position to argue the point. My mother heard the news first, and she was fairly furious, but she would not raise her voice or her hand to me until consulting with my father. It was Dad who usually handed out the punishments in our household, and here he could only offer his customary sentence: "Tomorrow, at six o'clock."

One of our favorite haunts was a small seaside shanty, l'Haricot (The Bean), that served the best fresh fish in the harbor. We could hardly afford to eat there as kids, but we loved to hang around the open-air café and take in the scene. It was the place to be, we all thought. The owner was a boisterous Italian fellow, who was quite a charming host, and the other White Wolves and I would feast on our own fresh catch before wandering by. There, we'd watch the owner make one of his trademark paintings as his customers dined. It was his signature move—and a terrific marketing ploy. He'd set up his easel and quickly paint the dinner plate, which on its own could have been considered a work of art. Or maybe he'd paint the diners enjoying their fine meal. When he was finished, he would offer up his painting for sale: If the customer resisted, he would lower the asking price; if he still couldn't make a sale, he would simply hang the portrait in some vacant spot on the wall and move on to the next table.

L'Haricot is still there. I make a point of visiting whenever I'm in the area, and my return trips are always a restorative balm. On one visit, well after my childhood, I was approached by a man who was eating at a nearby table. It turned out to be a fortuitous encounter.

"Are you the son of Cousteau?" he asked.

"I am," I said.

"Well then," he began, "you might be interested to know that your father and I have a good friend in common." He then mentioned the name of a man who had been instrumental in developing the Squale mask—one of the very first commercially produced diving masks. For many years, coinciding with the first diving experiments of my father and les mousquemers, Squale was the leading manufacturer of snorkeling equipment. As my father's adventures became more and more well known, the company consulted with him on new types of equipment. They even went on to produce diving suits for the French Navy, a more sophisticated, streamlined version of the patched-together rubber suits first designed by les mousquemers, although eventually the company was sold and the brand name fell out of favor. Now that I had run into this fellow at l'Haricot, it occurred to me that I hadn't seen or worn a Squale product for many years.

A side note: Those masks had a tendency to fog. To prevent this, we learned to rub a raw potato against the glass. We learned this from my father, who had discovered the trick by some chance or other. Unfortunately, we didn't always have a raw potato at hand when we were at sea, so we figured out that human saliva worked just as well. That's why, even today, you see divers all over the world, enjoying the most sophisticated equipment, still spitting into their masks before fitting them in place. It's the best, simplest precaution—and it costs far less than any of the products on the market designed to do the same thing. Plus, it certainly beats a raw potato.

Back to the friend of a friend of my father at l'Haricot, the seaside café. He said, "For many years, I was a shareholder of the company. I still have my Squale stock certificates, but they're now worthless. If you'd like, I can send them to you, as a keepsake."

"That would be so nice of you," I replied, never thinking the man would follow up on his offer.

I had become accustomed to meeting distant acquaintances or associates of my now-famous father, who would promise to share an old photograph or piece of personal memorabilia, but then I would never hear from them again. This was just as well, I always thought, and here I exchanged another few pleasantries with this man and thanked him for his kindness and assumed that would be the end of it. However, a few days later, I received a large envelope containing six stock certificates—and I was overjoyed. To the man at l'Haricot and to former shareholders, the certificates were painfully worthless, but to someone who had grown up with Squale equipment, whose father had tested and advised on some of the company's early designs, they were a welcome treasure. And they were beautiful, too, printed on fine parchment paper and decorated with colorful drawings—real works of art.

I share this memory here, in this context, for the way it reinforces the life I lived as a boy, enjoying many fine adventures in and around the sea, and also for the way it introduces the idea that my father's exploits and experiments would soon make a far bigger splash than any of us could have foreseen in those first years after the war in the waters around Toulon.

5

SANARY

In the summer of 1943, my father calculated that he made more than 500 dives with his new aqualung.

Two years later, he and his colleagues could count another few thousand more.

Two years after that, I imagine they had all logged more hours underwater, at greater depths, than any other group of divers on the planet.

By the time the war was over and relative peace had returned to the south of France, my father and his fellow mousquemers were clearly at the forefront of the diving community. That there was a diving "community" at all had much to do with their enthusiastic and often ingenious efforts. Certainly, they had done as much as any other group to introduce diving as a sport and pastime in the

Mediterranean—and beyond. Gone were the days when they imagined themselves as ancient pearl divers, able to hold their breath for more than three minutes and to plunge effortlessly to depths of 65 feet and more. Gone, too, were the limited and ultimately frustrating explorations in underwater science fiction, when they rejected the hard-hat equipment that navy divers had been forced to wear to service the keels of their ships, to scout for enemy mines, or to examine the wreckage of scuttled fleets.

In their place was a new spirit of adventure and discovery, and an opportunity to explore the oceans in a way that human beings had never done before. The three musketeers of the sea cleaved for dear life to that opportunity. Very quickly, they had come to the realization that it was the thrill and triumph that found them on the ocean floor that made life dear. It gave them purpose and meaning. Does this make sense? No, I suppose it does not, and yet in just a short time, the lives of these bold, adventurous fellows had been changed in such a way that they could no longer recognize the men they had been or the lives they had been living above the surface of the sea, before chancing to discover the wonders of the deep.

But here is an amazing thing—my mother had seen what my dad would become. All along, she was braced for this transformation. She had closed her eyes and pictured it, and then she waited for it—patiently, purposefully—until finally my father could stand before her as the man she had always known he would be. The dreams he didn't even know he had, and the dreams my mother had for him and for the two of them as a couple and for the four of us as a family were now close enough to touch. He hadn't even known to reach for them, and here they were. His new spirit of discovery energized him, this was true, but my mother had always imagined him so. In her bones, she never expected to be the left-behind wife of a navy captain. She knew the time would come when her husband would return to her, fully formed and in some ways reshaped by his experiences during the war and in many ways more reshaped by the sideline pursuits that now seemed to occupy his full and foremost attention.

I think back to how things were just after the war. I remember it as an idyllic time of great excitement and expectation—although there were also some difficulties as my brother and I adjusted to another important change. We were finally obliged to go to school, as I will soon explain. But at home, there was a new and renewed sense of normalcy. We were finally all together, in an open-ended sort of way. Dad would return after a day at the base or following one of many haphazard diving expeditions undertaken with the GERS, and my mother would look at him as if to say, "Ah, JYC, I knew you would come back to me before long."

That is the crux of it: My mother realized one day that the life she'd been longing for had finally arrived—and she took to it fully. She took the rest of us right with her. Our base of operations had been the Villa Reine, where we slipped easily into our new routines. It was a happy, happy time in the Cousteau household. All along, my brother Philippe and I had been enjoying childhoods like no other. Each day, Dad would drive off in his small green sports car, which he had fondly christened Ballilette. It was a convertible two-seater with a rumble seat barely big enough for me and my brother. Each night, he would return, and the car would fairly announce his arrival, such was the thick, rumbling noise of the engine as it careened up the rugged coast. Where he went in the morning was still a matter of uncertainty. Some days, he drove all the way up to Paris—along the famous ramshackle road known as Nationale 7, back before the development of the national highway system. Some days, he joined his friends on an impromptu diving mission, perhaps to explore the wreckage of a naval vessel or to inspect the harbor for mines. Other days, he worked on various experiments in diving physiology, putting his theories to the test with the help of navy divers. It was a job, but it was a job unlike any other, because he was not tethered to a desk or to routine.

During this period, Dad continued to film and explore shipwrecks for the navy, as well as record these expeditions for his own use. He produced his second film—*Epaves,* or, *Shipwrecks*—which debuted to great fanfare at the first-ever Cannes International Film Festival. Of course, he and his friends had been working on this movie for many months, and the timing of the festival happily coincided with its completion. In September 1946, Le Festival International de Cannes quite famously billed itself as the first international cultural event in postwar Europe, and it was just that. Dad and his friends joined a lineup of filmmaking luminaries that included Roberto Rossellini, David Lean, and Billy Wilder in presenting their new work. We traveled as a family to Cannes, together with Frédéric Dumas and Philippe Tailliez and their families and close friends, and I can still remember the joyful reaction of the audience as the film played on the giant screen for the first time. No one had ever seen anything quite like the astounding footage on display. Indeed, no one had ever imagined that a group of men could swim so freely in the sea, among such exotic creatures, or that a camera could record their movements so magnificently. It was a grand, celebratory success, and my father soaked it in.

The one wrinkle to this exciting time in our lives following the war: Philippe and I could not be kept from our studies indefinitely, and so we were finally enrolled in a formal school setting. Our endless string of "vacation" days came to a sudden end. Amazingly and tellingly, neither one of us had any long-term difficulty joining our age-appropriate classes, despite the running start of our classmates. However, it did take several months—perhaps even longer—for us to reach a kind of level playing field. Neither of us could read terribly well to start, and looking back, I suspect this was most troubling as it related to me because I was the oldest. (At seven, I should have certainly been reading.) Still, we had a facility for language and numbers, so we had a good foundation in place for when we finally got around to it. The trick came in finally getting around to it.

Up until this time, our education had consisted primarily of improvised sessions with our parents, who were both endlessly curious and eager to pass some of that curiosity on to their sons. We were "homeschooled" only in the sense that what little formal education we did manage to receive seemed to find us at home or at our parents' sides. On any given night, for example, my father might have roused me from bed several hours after I had fallen asleep. "Wake up, Jean-Mine," he might have said, shaking me awake. "It is time to learn about the stars."

Then he'd take me by the hand and collect Philippe in his arms, and outside we would go in our bare feet to look toward the dark blue sea, the stones of our patio still warm from the day's sun. Our eyes took a moment to adjust to the clear night sky. Dad would point a finger into the air, and I would follow it and listen. "The evening star is always the first to come up," he might say, in a soft, magical voice that seemed to float toward me on the still night air, and then he'd follow it with a story. I remember feeling enveloped by the depth of his knowledge, the conviction of his storytelling, as he spun for me and my brother a fantastic myth or legend and tied it in to what we were seeing. Or he'd point out the different constellations, which, on a subsequent night, I'd have to point out right back to him.

The world was our classroom—although most of our studies were based in and around the sea. This was both by design, I always thought, and happy accident. We learned what we needed to know to survive and thrive as intrepid outdoorsmen and adventurers. (We learned about the stars, so we could look to them for navigational help when we were sailing the seas.) We learned what was interesting to my parents, what they thought might be interesting to us, or what might prove useful at some future date. My father remembered the rote, standardized education he had received as a boy, often at some of the finest schools in France, and it was a steady and constant regret how much time he believed he had wasted in the pursuit of useless information. He could not abide a system of education that pressed upon children a series of subjects that had no resonance or meaning. To his thinking, if students have no interest in a subject, they will forget whatever you manage to teach them almost as soon as the assignment is completed.

We went first to the public school in Sanary in October 1945. As I have indicated, I struggled at first—not for long, mind you, and yet a difficulty can seem interminable when you are in the middle of it. That is how I remember my introduction to formal schooling, as one frustration after another. For one thing, I was considered an unruly, undisciplined child, but I did not know that I was unruly or undisciplined. I did not know what it was to be obedient. It helps to understand that we had no real rules in our house—none, anyway, since our grandmother Mutti had moved away with Jean-Philippe and Françoise. She had been the disciplinarian in our patched-together family. My parents were certainly not strict, and my brother and I had always been encouraged to explore any reasonable distraction. Here in the classroom, though, we were expected to sit dutifully at our desks and listen to the lessons and never once bound to the window to consider a brilliant blue sky or a soaring bird or an approaching storm.

Dad held out no great hope for our academic achievement, at least not in our first year of school, but even he began to worry as the months dragged on without any real progress in our report cards. A part of him still believed it was better for children—better for his children, anyway—to run in the hills and splash around in the sea. My mother believed this as well. But now there was no denying that our inability to read, for example, was depriving us of much of the joy and pleasure of discovery. We had fallen behind and were now in danger of falling behind even further. Without meaning to, and without realizing it, we had been handicapped in our pursuit of knowledge, and now that we had a dismal public school record to show for it, my parents began to worry that they had made a terrible mistake. My father worried about this most of all, because he had been the most vocal about his opposition to formal schooling. Therefore, he started asking his friends and colleagues about the educational experiences of their children—a measure of progress, I suppose you could say, although the manner of his inquiry left a great many people scratching their heads.

"You have children about the same age?" Dad might begin one of these inquiries, once he had established some sort of connection in this area.

"Yes," would come the reply.

"Do they go to school?" Dad would ask, never once considering the absurdity of his question. Of course these other children went to school. It was the law as well as the custom. It was like asking these parents if their children ate breakfast.

Soon, Dad had collected enough responses to learn that many of his colleagues were sending their children to a private school in Toulon run by the Marist brothers, a Catholic order of priests dedicated to raising good Christians and good citizens. So he and my mother made hasty plans to do the same. Off we went to Toulon, where we had to dress in uniforms that resembled my father's navy attire—and to pray to God in a manner we had not experienced in our anything-goes environment. Dad would drive us to school on his way to the base each morning, but we would have to take the bus back home. On those long return trips to Sanary, when I was supposed to be getting started on my arduous pile of homework, I would often fall into daydreaming of those days when Philippe and I had only to follow a school of brightly colored fish as they moved in unison through the turquoise waters near our home.

Oh, how I missed that windswept time.

Had I been shortchanged by my parents' unconventional approach to my schooling? I did not believe so at the time, and I do not believe so now. As a child, I felt momentarily cheated that those unconventional lessons were now over, but I also realized that I had learned a great deal in my extended "pre-school" career. Mostly, I'd discovered nature. I'd learned to feel its poetry, its splendor. And, keenly, I'd learned to approach the unknown with an open mind and an open heart and to always be prepared to be astonished—essential qualities, wouldn't you agree?

Meanwhile, life went on—not quite as before, but ever onward. We got a dog, a German shepherd we named Merou. He was a good dog with the bad habit of biting anyone whose face he didn't like—or, anyone who deigned to wear a uniform. Apparently, that was the province of the master of the house and the two Marist schoolboys. To be completely accurate, Merou was more of a nibbler than a biter. He never really hurt anyone, but he barked and nipped at them in such a way that the postman, other navy officers, and an occasional meter reader might have had their trousers torn slightly. When that happened, my parents made sure to replace the pants or the shirt or whatever garment that our dog had damaged. Eventually, they were spending as much money on other people's clothing as they were on ours, so they found another home for Merou.

We were helped along in our studies by a private tutor—an unusual setup in those days, particularly in a seaside outpost like Sanary, but now that they had finally gotten around to sending us to school, our parents were anxious for us to regain any lost ground. Somehow, they located and retained an eager young man to help us with our numbers and our letters. He lived in a small room over our garage, so he was almost always available to help with our lessons. This did not mean, however, that my father entirely gave up on his own efforts to teach us. He continued to bound into our bedroom at night to share some new discovery or other. Or, he would read to us from one of his favorite books— Kipling's *The Jungle Book,* for example—which very quickly would become one of our favorites as well. On some evenings, when the moon was full and he'd grown tired of the sound of his own voice, he would throw off our covers and lead us outside for another lesson on the stars.

It was altogether a time of enchantment, despite the new drudgery of school and homework, and soon our household would undergo another exciting transformation—one that found us by virtue of proximity. By great good fortune, the Villa Reine was located near a stone quarry. At first, this did not seem such a fortuitous turn because we could hear the blasts and the pounding all day long. It became like an unfortunate piece of background music, played over and over. But then, my parents learned that the quarry was about to be shut down, once it had been dynamited down to next to nothing, and they started to hear what was left of the racket like a symphony. You see, they smartly reasoned that nobody would want to own a parcel of land that had been blasted to oblivion, or one that had caused so much distress among its neighbors, so they thought they might be able to purchase it for a bargain-basement price. It featured, they said, the most breathtaking views in the region, and they spent many joyful hours contemplating the dream house they could build there. They even took me and my brother out for a look one afternoon and encouraged us to imagine what a house might look like if we built one on that spot.

Sure enough, their speculations proved accurate, and they were indeed able to purchase the land on my father's modest salary with the help of a small, no-interest loan from the French government, whereupon they set about building our new home. It truly was a magnificent setting. Looking out onto the water from that blasted-out quarry was like being at the prow of a ship, sailing right through the waves and past the big rolling whitecaps that dotted the coast in rough weather. My father had a name for those big foam crests—"sheep." Philippe and I made a game out of it. We imagined that those cresting waves were actually a herd of sheep, crowding together and being steered toward the shore. The name gave rise to a nickname my mother would carry for the rest of her life—La Bergère, or the "sheep keeper." The name was particularly appropriate in translation, because you could also hear it as "ship keeper," and in future years, she would become just that. She would be the true captain of the *Calypso*, my father's famous ship, just as she would hold sway over the sea itself, mastering the "sheep," or waves, as they escorted us on our travels.

Construction began almost immediately, and Philippe and I made a game of that, too. We got up early each morning to monitor the builders' progress and looked forward to our return from school each afternoon to see what new developments the day had brought. Together, we took turns wondering how our finished house might look and how our lives might be different once we were living there. Day by day, our new house seemed to rise from that wreckage of a site, although it didn't "rise" too terribly high because the house was a one-story affair. The idea of the design was that each bedroom would overlook the sea, so it really was like being on a ship.

My father enjoyed naming things. He took the job seriously. He believed that in naming a person, a house, a boat, or a funny cluster of whitecap waves, you allowed that thing to take on a life of its own. You gave it more than a mere identity; you gave it a personality. Early on, he named our new house "Baobab," like the tree, and it wasn't quite finished when we moved in. For weeks and weeks, construction workers were underfoot, but even in its raw, unfinished state it was an enthralling place, and it remains so more than 60 years later. I continue to call Villa Baobab home. I don't live there full-time, unfortunately, but I return each year for however long I can manage. The longer I stay, the better, and each time I return, I race to my old bedroom window and smile at the way Dad's "sheep" continue to crowd one another on the crests of the waves.

Very quickly, Villa Baobab became my mother's house. She filled it with her joy, her generosity, and her personality. I can still feel her presence. One of the big reasons for this sentiment is a lovely portrait of her that graces our entryway. You see it the moment you step through the front door. In it, she is the young woman I remember from the time we moved to this very house, wearing a simple, flowery dress, and flashing a warm, welcoming smile. Our future as a family seems etched on her beautiful face, and in her eyes, you can see hope and astonishment and all those wonderful things. It is as if she knew our lives were about to begin anew, in a fundamental way, and they would begin here, in our grand new home built on a seaside quarry.

It's one of my favorite pictures of my mother, but it is not my very favorite. That honor goes to a portrait I have not seen in many years—painted by my father, of all people—and I must pause here to share that story. You see, in addition to his many other talents, Dad painted in his spare time, which unfortunately meant that he painted hardly at all. Nevertheless, he had a flair for it, but usually not the time. Every time he started a new painting, he would eventually set the canvas aside before finishing it. He would be called away on some urgent matter or other and lose whatever momentum or inspiration he might have had in getting started. And so, as far as I ever knew, he only completed two paintings in his entire career as an artist. Both were portraits. The first was awful, although I had to admire his choice of subject: Christ himself. For a man who was only nominally religious, it was an odd selection, and I'm afraid JYC did a dreadful job of it. What I can only assume was meant to be His face appears to me like a rock, surrounded by bushes—which in turn I can only assume were intended as a crown of thorns. It's a disturbing likeness, although in truth, it's no likeness at all. For some reason, Dad gave Jesus one blue eye and one brown one, and when I first saw the finished painting, I wondered if this was just an amateurish or careless mistake or if perhaps my father was investing this visage with a deeper meaning. Maybe it was his way of expressing hope and despair in the same face—who can say?

The second completed piece was a portrait of my mother, and it was truly, truly magnificent. She appeared on Dad's canvas just as she was in life—filled with energy, vitality, warmth. Her élan is on full display, and when I first saw this painting, I wondered if the source of inspiration might also be the source of talent, because the two paintings certainly did not appear to have come from the brush of the same artist.

My father saw how much I admired this portrait and offered it to me as a present—but it was to be a present placed on layaway. He wanted to keep it at his home in Monaco, where he could still enjoy it. He said, "When I am gone, Jean-Michel, you shall have it."

I never forgot that gift, but I could not locate the painting when I was going through his personal belongings and settling Dad's estate several years later. I looked all over, but, inexplicably, the painting had vanished with the years. No one had any idea of its whereabouts. I was heartbroken, and the heartbreak comes racing back every time I step through the front door at the Villa Baobab. Why? Well, I see the surviving portrait hanging in its customary place in the entryway, and my mind flashes back to the lost tribute from my father's brush. But then my heart is lifted, because when I shake the sense of loss, I can see my mother herself—gracing this home with her beautiful presence. It is a sadness and a sweetness, all mixed together, and I realize I have returned to the one place we were truly alive and thriving as a young family—the point of departure for a lifetime of exploration and adventure that would soon find us all at this very spot.

WAITING FOR
CALYPSO

My father and his friends were becoming quite famous in our little corner of the world. Their triumph at Cannes, where the film *Epaves (Shipwrecks)* was awarded a special prize, made international headlines, and the Cousteau name was soon synonymous with underwater exploration. Les mousquemers had achieved a certain level of notoriety prior to the film festival, but it had been limited to navy and diving circles, and had hardly reached beyond the south of France; now they had slipped into the mainstream of public attention, and we were swept along with it on a wild ride that continues to this day in many respects.

I cannot say for certain how my father emerged as the best known of the group, for surely each man contributed equally to their adventures and discoveries. Looking back, one might think Frédéric Dumas a more likely candidate to become the "star" mousquemer, because he was the most gifted and accomplished diver, but Dad was the most gregarious of the three, and I suppose he took to the limelight most naturally. Another argument favoring my father: After spending long stretches of his childhood in the United States, his English was quite good, which cast him as a kind of default spokesman for the group when facing the international media. Finally, and significantly, it was his passion and aptitude for filmmaking that had spurred their efforts in this area, so it was probably fitting that he was out in front.

It was his "vision," after all.

In any case, people started to take note. Already, the navy had been making good and important use of the group's talents in this area, and the by-now burgeoning diving community had acknowledged Dad's important contributions and those of his cohorts. In and around Sanary-sur-Mer, I was accustomed to being recognized in my father's company. I quite enjoyed the extra measure of attention. When people did not know me by name, they knew me as the son of Cousteau—such was the reach of his reputation. And so my brother Philippe and I moved about in JYC's ever-lengthening shadow. However, this had been the case for some time, and now that my father's fame was beginning to spread beyond the Mediterranean, we did not notice anything different, at least not at first.

To us, in our world, life was much the same. We continued to dive and explore as a family. We continued to go to school and pursue our various adventures. Dad and his friends continued to disappear for weeks at a time on expeditions. And, as ever, we continued to regain our rhythm as a family upon his return and thrill to whatever discoveries he thought to share, picking up our routines wherever we had left them. The only difference, from our youthful perspective, was that we were now out on our own, heading off to school each day, and the new people we met as our world expanded would seem to know us before even meeting us. It might have been disconcerting, to move about as the "son of Cousteau," but it was only a gradual shift from the way things had been, so we hardly noticed.

Perhaps the one discernible change to our circumstance was in my parents' social circle. Very quickly, they went from mixing with an odd collection of naval officers and local divers and fishermen to entertaining all manner of French society. Artists, professors, scientists, government officials—everybody seemed to want to mingle with these intrepid undersea explorers, and my mother was a good and gracious hostess, staging fine dinner parties and receptions for this new assemblage of friends and associates. She took to the role as if she had been preparing for it all along.

I remember one grand gathering in particular, in the backyard of the Villa Baobab, when a roguish Dumas thought it would be amusing to treat our swimming pool with a fluorescent chemical that would turn the water a brilliant emerald green. My mother was in on Didi's plan, I learned later. (Over the years, La Bergère would prove to be an inveterate teaser, as prone to frivolous mischief as any other member of Dad's boisterous crew, so her actions would prove to be very much in character.) Here she thought it would be great fun for her guests to gather around the pool on a fine autumn evening as the water in the pool changed color. What she hadn't counted on, though, was Didi's impulsive decision to dive into the pool, encouraging the other guests to join him—fully clothed, of course. And what Didi hadn't counted on was the effect of the fluorescent chemical, which quickly turned his skin the same shade of emerald green.

One by one, my parents' new society friends jumped into the water. (Many, I'm sure, were gently pushed.) None had any reason to suspect that the dye would ruin their fine clothing, but it surely did, and this was especially problematic for one high-society guest, who had been standing by the side of the pool dressed in a splendid fur coat.

My poor mother didn't have much patience for this excessively elegant woman before she fell into the pool and destroyed her fine white coat—and now that she had done just that, my mother had even less. Truth be told, Simone Cousteau quite preferred the roughhousing and merrymaking of her husband's familiar acquaintances to the proper, snobbish behavior of the society crowd, and this fur-coated woman had fit herself quite haughtily into the latter category. For years, my mother would remember this backyard splash as a highlight of a time when our lives were about to change—and as a prophetic symbol of that change. She'd tell the story and punctuate it with a deep, throaty laugh, never much caring that the entitled woman had left in an ostentatious fit of outrage, followed by her equally outraged husband, slamming the house door behind them. My parents didn't speak to the offended couple for 30 years, but the stalemate only added to my mother's delight in sharing the story.

"What kind of idiot stands at the edge of a swimming pool draped in a white fur coat?" my mother would always say. "It was a party of drunken sailors. What did she *think* was going to happen?"

Eventually, the "drunken sailors" of my father's acquaintance gave way to a more refined cast of characters. Indeed, many of Dad's new contacts led him directly and indirectly to his next adventures. One such fellow was Auguste Piccard, the noted Swiss physicist who had invented a high-altitude balloon gondola that he was now looking to adapt for underwater use. To this end, he had designed a bubble-shape steel capsule that would maintain normal air pressure when it was lowered into the ocean, allowing divers and aspiring oceanographers like my father to record their activities at impossible depths. Piccard and his team had been working with the French Navy on their designs and had conducted a series of unmanned experiments when he sought out my father. The two men got along quite well and enjoyed sharing their ideas and theories. They also enjoyed playing vigorous games of tennis, and because I was an accomplished young player, I was often asked to join them. Even at ten years old, I could compete with my father and his friends, and in his own way, Dad enjoyed dragging me to the court and showing off my skills in this area almost as much as I did.

For my part, I enjoyed the interplay between Dad and Piccard as much as I did the tennis. It felt to me like I was being let in on a profound moment. Professor Piccard became a great friend of the family, and I loved hearing about his various experiments. He was quite ingenious—and, quite the showman. I wasn't fully aware of how famous he was at the time, but his pioneering work in the stratosphere had captured worldwide attention. People were eager to see what he might attempt next. I was young, but already I had an appreciation for adventure and discovery, so I came to admire Professor Piccard enormously and found myself looking forward to his visits. I listened with great interest as he discussed plans for his submersible vehicle, which he called a bathyscaphe. He was working closely with his son, Jacques, who would go on to his own fine career as an oceanographer and explorer, and he seemed to take particular delight in discussing their collaboration.

After one of Piccard's animated discussions about his work with his son, Dad would sidle up to me and say, "Someday, Jean-Mine, you and I will explore the oceans together."

It was something to look forward to.

In all, Piccard and his son would build three bathyscaphes, and in 1948 my father was set to participate in the first exploratory mission aboard the first of these vehicles—off the western coast of Africa, along the islands of Cape Verde. This was discussed in great detail one afternoon over tennis. Dad was excited at the prospect, but then we returned to our tennis, and his plans were nearly dashed. He lunged to reach one of my passing shots and took a bad fall, breaking his ankle in the process, but he accompanied Piccard anyway—with his foot in a cast.

My mother was not happy with my father's plans to join Piccard on his expedition. In fact, she was briefly overjoyed to learn of his injury, because she thought at first it might keep him from making the trip. She did not trust the bathyscaphe to keep her husband safe at depths never before experienced by man, and she did not trust her fearless husband to exercise good judgment when it came to assessing the risk of the endeavor. Dad could not contain his excitement, however, and he would not allow the small inconvenience of a plaster cast to keep him from his adventure. So, in October 1948, he sailed to Senegal with Tailliez, Dumas, and others of their team to join Piccard on a freighter, where the bathyscaphe awaited.

Les mousquemers had never seen the contraption before, and they marveled at its construction. The photographs and sketches did not do it justice, they said. Even more, they were awed by the thought of what they could accomplish with the bathyscaphe if it was proven successful. Piccard had even designed a metal claw, which extended from the vehicle and allowed for the collection of specimens from the ocean floor.

It had been only a few years since Dad and Emile Gagnan had perfected their aqualung design, but already the most determined, most experienced divers could see the limitations of the human body. We were air-breathing animals, after all, built to operate at atmospheric pressure. The new technology could only take us so far.

Without a pressurized chamber like Piccard's bathyscaphe to take us to the ocean floor, we would never be able to safely descend below a certain depth and explore the ocean floor like true fish, and so these developments were particularly thrilling to underwater explorers.

Regrettably, Dad did not join Piccard on the bathyscaphe's first and only manned descent. That honor went to another member of Dad's team, the French naturalist Theodore Monod. But my father was hopeful that he would have a chance on a subsequent dive later that same day, or perhaps the next morning. That chance never came, however. After a thrilling trial run, with Piccard and Monod in the spot JYC would always believe was meant for him, the group deployed the bathyscaphe on an exploratory test over an undersea canyon. This time, something went horribly wrong, at a record-breaking depth of over 4,600 feet; one of the chambers became bent and was rendered useless. The rest of the mission had to be scrapped. Piccard and his son went back to work perfecting their design, while JYC and his friends returned to Toulon.

All was not lost, however. Dad's cameras captured some spectacular footage of the manned descent, and the expedition in turn captured headlines all over the world. Once again, the Cousteau name was out in front—quite curiously, in this instance. No, JYC had not been directly involved in Piccard's mission, and indeed most of the spotlight had fallen justifiably on Piccard and his team. You could even argue that JYC had not participated in the manned descent at all, and yet he was connected to it in a high-profile way. He had consulted on the design and in the planning. Moreover, he was there. On top of the splash he and his friends had made at the film festival in Cannes, justifiably or not, the aborted bathyscaphe experiments placed Jacques-Yves Cousteau at the forefront of the world's diving community and cast him as one of the foremost explorers of the world's oceans.

There followed a rush of new attention. *Life* magazine published several pages of Dad's still photographs, together with his compelling commentary. His comings and goings were tracked by the press, and his various salvage, archaeological, and minesweeping missions for the GERS aboard the sloop *L'Eli Monnier* were frequently given front-page attention. More and more, he was asked to comment on some underwater exploration or another, whether or not he was directly or even remotely involved. I would not go so far as to suggest that my father had become a celebrity, but where he once did his work in relative obscurity, he now moved about in the public eye. His accomplishments were being celebrated far and wide. People knew his name, his face, his reputation. It got to where my grandfather, Daniel Cousteau, began fielding offers on JYC's behalf from various organizations. My grandfather had been working in the employ of successful international businessmen. He had developed important contacts all over the world. As an attorney, with experience in a variety of fields, he was well positioned to guide my father through these next paces and to negotiate on his behalf, so Dad asked him to run interference on these matters, allowing him to concentrate on his expedition work.

All over the world, all of a sudden, people wanted to hear Dad speak of his underwater adventures or to screen one of his films. Advertisers wanted him to lend the Cousteau name and allure to their products. Dignitaries wanted him to grace their important events with his presence. Even Hollywood studios started calling, with funding for his next film. Somewhere in this confluence of opportunities was our new reality: Dad had become an overnight sensation—and even though the "overnight" had been more than a decade in the making, there was no denying the sensation.

This shift coincided with another important change in our lives— namely, the procurement of a vessel my father could call his own, a vessel that would become his base of operations. For nearly four years, he had sailed *L'Eli Monnier* for the navy, but JYC believed the sloop unsuitable for his future needs. Even as a boy, I could see his point. Mostly, it was too small, and he said as much to the admiral in charge of the GERS team. To Dad's disappointment, the admiral denied his request for a bigger vessel, telling him he needed to concentrate on advancing his naval career if he wanted to command a bigger ship. The one would follow from the other. JYC made a similar appeal to another admiral but was shot down a second time, so he began to think seriously about resigning from the navy and focusing his full and foremost attention on his underwater explorations. There was money to be made from movies and books and lectures, he felt certain. He was also certain that there was money to be made with the successful marketing of the aqualung device. The only uncertain piece was how exactly to go about it.

Then, in the spring of 1950, he had a vision. Or, at least, he liked to recall that he had a vision, but it's possible that what became a vision in the retelling was in reality a chance encounter. Either way, he'd found his ship. In the first version of the story, the magical version that suggests that the *Calypso* appeared before him like something out of a storybook, Dad was out for a stroll in the harbor village of La Valette-du-Var, on the island of Malta. La Valette is a charming village that even today feels old and new all at once, rich with history and at the same time utterly contemporary and altogether vibrant, although in 1950, the large naval ships in the port were a bleak reminder that the island was still under British rule.

In this very harbor, my father took to saying that the *Calypso* appeared before him for the first time. He fell immediately under her spell. He described the ship to his collaborator Yves Paccalet in the following terms: "With her half-white, half-black wooden planking, anchored among the fishing boats and battleships, she

113

enchanted me immediately. I want her. I shall have her. I read her name on the hull: *Calypso C.* At that moment, I realized that I will command her and that I will sail her to the end of the world. . . ."

It is a romantic vision, yes? And yet there is another account that seems far more reliable. I heard my father tell this complementary version many times as well and came to suspect that the truth rested in the balance of these two stories. In the second version, my father was visiting Auron, a ski resort in the south of France, and he was at dinner with friends, talking about his dream of exploring the world's oceans. He expressed his frustration with *L'Elie Monnier* and told of the difficulties he was having requisitioning a more suitable vessel from the French Navy. JYC was a wonderful dreamer, a wonderful storyteller, a wonderful dinner companion—and from the accounts of those present, he was his usual effervescent self on this night. When my father stood to leave, he was approached by a gentleman who had been seated at the next table. The man appeared to have been caught in the swirl of my father's enthusiasm. He said, "Forgive me, but I overheard your conversation. I might be in a position to offer you a vessel to help you achieve your dream."

The gentleman was Loel Guinness, of the Guinness beer and banking family, and he had a ship on the island of Malta that he thought might be suitable—a wooden-hulled minesweeper that had been built in the United States out of Oregon pine and had been commissioned by the British Royal Navy in 1942, after which it was stationed in the Mediterranean for the balance of World War II. Following the war, the ship was sold to a private operator and used as a ferry between the islands of Malta and Gozo. There, she was rechristened *Calypso,* after the mythic nymph of Homer's *Odyssey,* who kept her companion Odysseus in a cave on an island that was said to have been inspired by Homer's visit to Gozo.

My father traveled to Malta to inspect the ship and found it to his liking. I have since concluded that this was where the second version of the story dovetailed with the first, in the harbor of La Valette. Yes, the *Calypso* appeared before my father and fairly announced herself to him, as he so quixotically described, but it was no chance encounter. Yes, it was a kind of love at first sight, but it did not catch him unawares. He had traveled to Malta for the express purpose of inspecting the ship, and to hear him tell it, he was seduced by what he saw. It felt to him like the ship was waiting for him, welcoming him, when in reality, he was the one who had been waiting to welcome a ship like *Calypso* into his life and work. He especially liked that *Calypso* had a wooden hull, which he believed would offer a more comfortable journey than the metal ships he sailed with the French Navy. In any case, he said, it would be much quieter—and, therefore, less likely to scare away the beautiful fish he was so intent on observing. I remember, upon his return, hearing my father marvel to friends that you could fit 11 automobiles on board. This, to him, was magnificent, although I could never imagine a scenario where we might need 11 cars out at sea.

In any case, my father reached an agreement with Guinness, who would only accept the token sum of one franc per year to lease the vessel. Furthermore, Dad could not disclose the name of the ship's owner. (It was only upon Guinness's death that my father spoke openly of his benefactor.) I never knew why it was stipulated that Guinness's generosity could not be named, only that it was. However, I did not question it. As a mere kindness, this man endowed my father with the wherewithal to realize his lifelong dream, so it was a profound and sustaining gift.

As part of the arrangement, my father visited a shipyard in the south of France, where Guinness kept *Calypso*'s sister ship, *Callisto*. My father had to outfit the ship at his own expense, but the work would be done by Guinness's crew. That was the arrangement, and it certainly made sense. Because *Calypso* would remain a part of the Guinness fleet after it no longer met my father's needs, it would not do for my father to refit the ship in a manner Guinness deemed inappropriate. The ship was redesigned as an expedition and observation vessel and featured a see-through "nose" below the waterline and a helicopter pad up above—it was difficult for me and my brother to reconcile our first impressions of the *Calypso*, when she was in some disrepair, with the refurbished ship we came to know and love.

Dad could not resist the charms of the nymph. His friends told him he was like a hopeless fool and rather like the legions of sailors who had fallen in love with their ships. Over the years, if you watched him carefully, you could see him treat the *Calypso* like a person—like a woman, of course. He would even talk to her in times of difficulty, beseeching her to take care of his crew or assuring her that he would take care of her. He started to think more and more about retiring from his military career. In his excellent memoir, *The Human, the Orchid, and the Octopus,* written with another longtime collaborator, Susan Schiefelbein, he wrote quite movingly of the professional crossroads he faced now that he was commanding the *Calypso*. "I see myself again dreaming that if I attempt the adventure of filmmaking and travel, I will cut myself off," he wrote. "That I will give up reality for a shadow, a certainty for a fantasy."

One of the great themes of my father's life, however, was that he had the unwavering support of his fellow mousquemers, as well as that of his wife, no matter the depths of his folly. He would not resign his naval commission for another few years, but here he was weighing his options. When others thought he was possibly mad to invest everything he had into the *Calypso* and set off on the uncertain waters of a full-time career as an oceanographer and filmmaker, Tailliez, Didi, and La Bergère stood steadfastly behind him. Indeed, my mother was the first to do so. "This doesn't surprise me," he reflected at length in *The Human, the Orchid, and the Octopus.* "She is with me all the way. She backs me with her love, morally and financially. Her family is more affluent than I. Simone helps me. She is, in fact, even more adventurous than I am. She wants to go far, beyond the edge of reason. She has the soul of a sailor."

The soul of a sailor? Beyond the edge of reason? Surely, these sentiments could be applied to each of my parents, and bookshelves are fairly filled with musings on the romantic, improbable quest of my father's life, but little has been written about my mother in this regard. What most people don't realize is that my father was helping my mother realize her dreams, just as she was helping him to realize his. La Bergère was a young woman in love, with an idealized vision of what it meant to live a life of the sea. Of course, she could never be an admiral like her father and grandfathers, but she could do the next best thing: She could sail the world with her husband—her "Loubi," as she called him—and in the process be a kind of catalyst for others to do the same. She had never wanted for anything else—and here it was.

On the *Calypso,* my mother was immediately at home, and she took immediate charge. In many ways, she was more captain than my father, an assertion to which even he would have readily agreed. In short order, the *Calypso* became our base of operations as a family. It was our heart, our center, our core. Moreover, it was our home.

Once, early on, when a reporter made the mistake of asking my mother where she lived, she answered resolutely, "Why, on the *Calypso,* of course!"

Philippe and I overheard that remark and others like it and realized with regret that the ship could not be our full-time home as well—not for the time being, anyway. We were school-age children, and this had to be considered. However, I don't think either one of my parents gave our continuing education any real consideration until the very last moment, whereupon the Melchior side of the family weighed in with a recommendation that they consider boarding school as an option. Specifically, my maternal grandparents told my mother to enroll us at Ecole des Roches, one of the best-known boarding schools in France. Their argument was plain: It had been good enough for our uncles and cousins, and so it was good enough for us. At the time, the school did not have quite the same reputation it enjoys today. Make no mistake: It was a fine institution, but it was not the very finest. Philippe and I would board with the sons of some of France's finest families, but here again, not the very finest. In this one respect, at least, it was a dramatic change from our experiences at the public school in Sanary or the Marist school in Toulon, where we were often seated next to boys from all walks of life, of every conceivable stripe and station.

Perhaps the biggest difference was that my brother and I no longer saw each other—at least not on any sort of regular basis. It was difficult enough to be separated from our parents, but now my brother and I were separated as well. Philippe was enrolled in the primary school, whereas I attended the secondary school. Les Roches was in the country, in Normandy, at some distance from Evreux, the nearest town. The school was situated on a sprawling, parklike campus, which was bisected by a set of railroad tracks that offered a kind of border that effectively separated the school into two distinct sections. The younger boys were on one side, and the older boys were on the other, and we weren't allowed to cross those tracks for any reason. Moreover, the culture of the school was such that we didn't want to. The older boys saw no need to mingle with the younger boys, who they treated like unwelcome pests.

This left me feeling alone and disconnected. By now, I was a good student, but even my studies were not enough to distract me. I missed my family—Philippe, my co-conspirator, most of all, and he was only across the train tracks. I missed the adventures we all shared and the endlessly hopeful sense that anything could happen at any time. And so I filled my days with a hopeless boyhood crush. I became enchanted with the young French actress Marina Vlady. I stole away to the cinema in Evreux at every opportunity, to see her in *Orage d'été,* by Jean Gerber, perhaps, or *Avant le déluge,* by André Cayatte. For good or ill, this became my one and abiding extracurricular pursuit; if I could have "majored" in Marina Vlady, I would have surely done so. Whenever one of her pictures was playing, I went to see it, over and over, such was my infatuation. There was a bus that reached to the outskirts of Evreux, and from there, I would run to the cinema in time for the first showing, and whenever I could manage it, I would stay in that darkened theater until the very last showing that allowed me to return to campus in time for curfew. Marina Vlady was all I could think about, and in this way, I passed virtually all of my free time at Les Roches, at least for the first few semesters.

I now realize that this one-sided love affair was meant to fill the spaces where my family had been and to help me pass the time until we could be together once more. In many ways, I enjoyed my time at Les Roches, but it was intolerable in many ways more, because just as my parents had been seduced by the *Calypso,* I had been seduced as well. I longed to be with them on the high seas, laughing with the crew they had hastily assembled before the ship's maiden voyage under Dad's stewardship—in June 1951.

When Philippe and I arrived for the dedication in June, I was astonished at the ship's transformation. The *Calypso* had undergone such a complete and thorough makeover that the broken-down minesweeper I remembered from my first inspection was nowhere in evidence. Indeed, I had spent many weeks with my father's crew helping with the refurbishment, but I had been away for several months by this point and now hardly recognized the vessel.

My parents had invited a great many friends and associates to sail with us on this first voyage. The well-known naval architect, André Mauric, was there with his brother Edmond, also an architect, who had designed our custom Villa Baobab home. One of Dad's filmmaker colleagues was also on board—Jacques Ertaud, whom we all called Jacky—together with Roger Gary, an industrialist from Marseille, and his well-heeled brother-in-law, the Marquis of Turenne. Personally, I was most delighted to see *Calypso's* chief mechanic—Octave Leandri, or Titi for short. Titi had been the first person Dad hired when he took over the ship, and I was drawn to him because, at 23 years old, he was only ten years older than I. We got along quite well and developed a fine friendship over the years, so much so that we continued to sail together on expeditions and to see each other long after the *Calypso* had been retired and my parents had been put to rest.

Back then, Titi was an exuberant, boisterous soul, eager to please the "pasha," as my father was called, and La Bergère. He loved to tell the story of how he was hired. The *Calypso* was docked in Antibes in the spring of 1951. Titi had just returned from Indochina, where he had spent two years as a marine mechanic, when he heard from his cousin that a naval lieutenant was looking for someone to help him fix an old tub. (Sadly, that is how the *Calypso* was known among the sailors of the region, insofar as she was known at all—as a rundown bucket of wood.) Titi dutifully went down to the dockyards to apply for the position and found Dad striding the deck of his new ship like a proud captain. The picture belied the scene, Titi always said. JYC was tall and slim and cut an elegant figure, whereas *Calypso* was clunky and dilapidated. Nevertheless, Dad bounded over to Titi and waved him onboard and told him to start up the engine.

Titi thought my father was crazy, and I could certainly understand that view. Here he had handed over the controls of his new vessel to someone he had barely met—but that was my father. He had a good feel for people, he always said, and here Dad must have felt something because Titi never left. He ended up staying on the *Calypso* night and day for the next several weeks, as she sailed from Antibes to the dockyards of La Seyne, near Toulon. Dad had chosen the location because he had to pass through La Seyne every evening on his way home to Sanary from the naval base, so he could check on the ship's progress. When the *Calypso* was finally ready to put out to sea, he handed Titi his final paycheck and made his good-byes.

"What are you going to do now?" the captain asked the mechanic.

"I don't know," Titi said. "I suppose I will look for work."

"Well, then," JYC said, "you will work for me."

And so it was decided. Titi himself had no say in the matter, but he had nothing better to do. Of course, he had no idea the *Calypso* would become famous. None of us had any idea, really—except perhaps for my father. The *Calypso* was only 138 feet long, from bow to stern. She was barely 30 feet wide. And she weighed just under 300 tons. But she was refashioned and refitted and ready to sail the world's oceans with Captain Cousteau at her helm.

Unfortunately, Philippe and I would not be along for the ride— not just yet. We returned to Les Roches following that inaugural voyage with heavy hearts and teary eyes. Up until this time, the long stretches away from my family had been difficult, but now they were intolerable. Now we had a brand-new taste of what we were missing. I found myself counting the days until our Christmas vacation, when we would join our parents once again. The pages of the calendar could not flip fast enough to soothe my wandering spirits. Already, the *Calypso* had begun to feel like home to me and my brother. We had taken some short voyages in the Mediterranean when my father was testing the ship, and we had each spent countless hours scrubbing the decks, cleaning the galley, painting the hull, and helping to make her seaworthy in what ways we could as boys. For the inaugural, we moved about on the refashioned ship like we had been born to it. But then we were cast aside, without ceremony, and returned to our fine boarding school in the Normandy countryside.

Soon, before *Calypso* could set sail on her first formal expedition and before Philippe and I could settle into our routines at school, our family was struck with the sad news that my grandmother Mutti had died. We had not spent much time with my grandmother since she had lived with us during the war, but my father had been in close and constant contact with his parents, especially since my grandfather had started seeing to his business affairs. Very often, when "Daddy" visited us in Sanary, Mutti stayed behind in Paris, where she was still caring for my cousins Jean-Pierre and Françoise. My uncle PAC had been imprisoned at a facility near Paris, and Mutti visited him regularly—dutifully, dotingly. Whatever his presumed "crimes" against the government, he was her son after all. My father felt the same way; PAC was his brother. Blood was thicker than politics, at least among us Cousteaus.

Indeed, it was during one of my grandmother's visits to see my uncle in prison that she suffered a fatal stroke. We received word as JYC was making final preparations for the *Calypso's* maiden voyage under his command. It was a devastating piece of news, and we all came together for the funeral. Uncle PAC could not join us, of course, due to his incarceration, but the rest of us were together for this solemn occasion, and I remember standing off to the side with my brother and cousins during the ceremony, wondering when we would all be together again. The war had split our family apart, and now Philippe and I were off in boarding school, my cousins were in Paris, and my parents were preparing to set sail on Dad's new ship. It felt to me just then that we were all embarking on an uncertain adventure.

Philippe and I retreated to Les Roches with heavy hearts—and yet with great hope as well, because we knew it would not be long before we joined our parents on board the *Calypso*. We looked forward longingly to our first expedition, our first grand adventure. It would become our routine over the years to join the *Calypso* whenever we were on holiday from boarding school—however, joining our parents on the ship was never easy. For this first adventure, it was nearly impossible. We were meant to meet the ship in the Red Sea that first Christmas, and we boys were tremendously excited. We were excited for the adventure and for the reunion, both. You see, my parents were now spending so much time on board the ship while we were spending so much time at Les Roches that already I was recognizing that we would only be together as a family—truly together—when we were at sea. The irony of this thought is not lost on me now, just as it was not lost on me then. Other families were quite figuratively at sea, struggling to find common ground and points of connection, whereas ours was most at home when we were out on the water, untethered by the constraints that bound the rest of the world, joined by the boundless expanse of the sea.

Over the years, my mother would spend even more time on board the ship than my father. One year, when the *Calypso* was in the Amazon, she spent ten consecutive months on the ship without sleeping one night ashore. The Cousteau men were in and out and on and off all the time, as were the other members of my father's intrepid crew, but my mother held fast. More and more, the nickname she carried among family and friends seemed especially appropriate: La Bergère—the sheep keeper. The ship keeper. Actually, the literal translation is "wife of the sheep keeper," but she was certainly in charge—of the whitecap waves that dotted the coastline beneath the *Calypso's* keel, and of the ship itself. Outwardly, for the rest of the world to see, my father was the apparent captain, but it was La Bergère who looked after the crew. It was La Bergère who collected the dispatches and telegrams from around the world and passed them along to our crew, for example—a disquieting task when the news from home was laced with tragedy and sadness. And as the ship's de facto nurse, La Bergère soothed the aches and pains and the bruises of our crewmembers.

It was my father who charted our course, but my mother determined our days.

A rendezvous had been arranged for us boys that first Christmas in the city of Jeddah, on the coast of Saudi Arabia. The *Calypso* was in the Red Sea at the time, but as so often happened in those days, Air France was shut down by a strike of some kind or another, and we could not leave Paris to begin our journey at the arranged time. We were forced to remain with our maternal grandparents for another short while, until the strike could be resolved. Nowadays, of course, we would take such a delay in stride because we could communicate with our parents directly via cellular phone or computer and perhaps make alternate arrangements, but it was not so easy back then. There was some trouble getting word to the ship about our delay, and then again once our travel plans were restored. When we finally did arrive, three days late, the ship had sailed. The *Calypso* would not wait for an undetermined period for the ship keepers' children to turn up on holiday. We would simply have to make other arrangements upon our arrival.

Philippe and I did not know any of this as we made our way to Saudi Arabia. We had no reason to think our parents would not be waiting for us and that our trip would not continue as planned. We went first to Beirut, where we had to change planes. We were traveling with some additional equipment—cameras and lenses and rolls and rolls of film, altogether quite a lot of gear to entrust to two young boys, but it was thought that it would be easier for us to carry the equipment than to ship it directly. We spent four hours waiting for our connecting plane—a DC-4 propeller plane, as I recall. We were the only French boys on board. It was a fascinating flight, really—our first long trip, and here we were, making it by ourselves. I remember a group of Arab men, huddled over a Sterno burner in the aisles, warming pieces of flatbread. It made an odd picture.

In Jeddah, we made our way through immigration, and it appeared our papers were in order, but the Saudi officials confiscated all of our equipment as we made our way through customs. We were just children, so we could not argue effectively for its return. I simply assumed that once we passed through customs, we would be met by one of our parents, and they would attend to the matter. But neither of our parents was there, and they had not sent anyone familiar to meet us. They had not abandoned us entirely, it turned out. There was a short man with a bushy mustache who had apparently been told to collect the two boys on the plane, but we did not know of these arrangements. He did not know what time our plane was scheduled to arrive but was told to look for it in the sky on approach—such was the haphazard informality of air travel in the Middle East in those days.

We had never seen this man before. I did not speak any Arabic, and this man could not manage French or English, so we struggled to make ourselves understood to each other.

I was terrified. I was in charge of my brother, and I did not want Philippe to know I was afraid. But somehow this man convinced me that my parents had sent him, and eventually we agreed to go with him. We had our personal bags but none of the equipment, and I knew it would be impossible to negotiate for its return, so we left. We drove through the desert for the longest time. It seemed like hours and hours. Outside our window, there was nothing to see but desert—and then, inexplicably, we came upon an amazing property. It was lush, like a hacienda, and beautifully appointed. It was the home of the French ambassador, we learned, and he received us warmly. Finally, we learned that my father had arranged for us to wait for him here, so we were greatly relieved. The ambassador's staff showed us to our room. We were exhausted. We had been traveling for many hours, and we slept until the next afternoon. While we slept, the ambassador made arrangements for the return of my father's equipment, so the situation had been righted by the time we awakened. I was responsible for my brother and for the equipment as well, and now I could relax because I had delivered all of my packages safely, even though my parents were still not here to receive us.

For the next few days, we waited out our parents' return with the ambassador and his family. The *Calypso* team took its time returning to port, because they were capturing some marvelous footage of sharks—and much of that footage would be presented to spectacular effect many years later on the very first episode of Dad's American television program, *The Undersea World of Jacques Cousteau,* in a segment entitled "Sharks." Meanwhile, our hosts on land were very kind to me and Philippe. They took us into the desert, where we saw a band of nomads catching grasshoppers, which they would hurriedly fry and eat. I had to try it. Always, I would try everything. I had a curious mind and an adventurous spirit (and a strong stomach, too), even as a small boy.

We visited Jeddah, the richest city in Saudi Arabia—and the capital of the country, at the time. There was tremendous excitement, tremendous haste. Philippe and I were plainly fascinated by this new environment. In a peculiar way, it was like being on one of my father's dives, as we floated through this unfamiliar sea of activity and tried to make sense of a world we could never truly know. At one point, we heard a big commotion and turned our heads to see a man being pulled from a building. People were shouting, and as we floated through the crowd for a closer look we could see that the man was about to have his hand cut off. Another man was standing over him with a large knife, making ready. Another few men were constraining the man in custody. And it was all happening right there on the sidewalk. Right in front of us. It was my first time in an Arab country. I could not understand how such a thing could be allowed to happen—in a public place, no less—but there it was. Apparently, the man had tried to rob a bank, and this was his punishment. If you reached for the money with both hands, they cut off both hands. If you reached with just one hand, as this man had done, they only cut off the offending hand.

I stood next to Philippe, transfixed by the scene. I was afraid to look, but I could not look away. And soon enough, we were moved along by the sea of people on that crowded sidewalk, on to our next adventure.

7

A Place
of My Own

I must admit, I was not a terribly good student. It would not be accurate to suggest that I struggled at Les Roches, but neither would it be fair to argue that I excelled. I got by, with no better than passing grades. Philippe, too. We were incorrigible, I'm afraid. Just as my father was becoming well known to the outside world for his many and fine adventures at sea, so, too, were my brother and I becoming well known to the boarding school faculty as the underachieving children of Cousteau. We were grudgingly accepted by our long-suffering teachers, who showed us a measure of tolerance we did not deserve on our own. They would say "our little birds from the islands are back"—derisively, whenever we returned a few days late from a school holiday, which as it happened was more often than not.

And yet I consider my time at boarding school a success, for it helped to nurture and nourish my ability to interact with all manner of people and develop a sense of discipline that could never quite find me at home for a variety of reasons. Teachers, school officials, classmates—I quickly became expert in navigating my way through these many relationships, even as I scrambled to keep up with my studies. Also, our frequent comings and goings forced me to adapt and fit myself into many new environments and to think quickly on my feet, so that after only a few years, I could look in the mirror and see a confident young man who could talk his way into and out of most any situation.

I almost couldn't recognize myself, but I liked the picture that came back in the reflection. My parents, too, seemed to take great pride in my development. As I have written, my father did not place the highest value on classroom learning, so he paid scant attention to my mediocre grades. They were meaningless markers. It was my emotional growth that was important to him and my mother—and an awakening of a boundless curiosity that seemed to be genetic, because it eventually surfaced in Philippe as well.

My "real" education during this period came on board the *Calypso* and in studying the habits and positions of my father and his cohorts. I was a student of human nature—and, soon enough, nature itself. It is often said that a student who struggles in school but excels in the ways of the world is possessed of street smarts instead of book smarts, although in my case, I would have added sea smarts to the profile. Gradually, in the manner of a wide-eyed adolescent and then an intractable teenager, I came to share my father's perspective on the oceans and the planet and the importance of conservation. Indeed, soon after this period when he started sailing the *Calypso,* Dad began to articulate his displeasure at the way man treated the environment.

Soon, I started echoing his views and reshaping them in my own particular way.

Now, all these years later, I have come to look on our planet as a magnificent symphony performed by a giant orchestra. Every plant, every animal, every living, breathing thing must play its part for the notes to come together in the most mellifluous way. Each is indispensable to the whole. The idea is to live in concert—even the language supports this view. Surely, if you place one hundred violins in an orchestra and remove just one, it will not be terribly missed. But if you remove a great many, one by one, without considering the whole, you will at one point remove just one, and it will be one too many. Suddenly, *bang!* Everything falls apart, and where there was once harmony, there is now discord. Where there was once music, there is now simply noise.

Such is the perspective of a lifetime in and around the world's oceans, but I believe it's helpful to note that these ideas took shape while I was still in boarding school, watching my father and his friends grow more and more concerned about our environment. Their concerns became mine, and I took them in over every vacation break from my so-called studies at Les Roches. In this, Philippe and I were cut a little differently than our fellow students. They usually joined their families in the Alps for skiing or on some Mediterranean shore for swimming and sunshine, but we always made for the *Calypso*. Wherever she was, we would find a way to join her, very often flying off toward some far-flung outpost with no clear idea who might meet us, or when, or where.

Still and all, we became quite adept at fitting ourselves into the rhythms of the ship and its crew, which had by now become a kind of extended family. From the moment she first sailed with my father, the *Calypso* was our first and foremost home. It was where I imagined myself when I closed my eyes and pictured my family.

Curiously, many of my father's crewmembers were only a few years older than Philippe and I, yet we addressed them with the more formal *vous,* while they in turn addressed us as *tu*—a holdover, perhaps, of a more rigid upbringing than our own. Very quickly, however, my father established an esprit de corps that hardly resembled the formality of his time in the navy or the rank and file you were likely to find on any other vessel. There was necessarily a certain measure of rank aboard the ship and a clear chain of command, but Dad would not be the type of captain to arbitrarily give orders without explaining his purpose or setting forth his goals. He wanted all hands on deck to fully embrace the spirit and wonder of each expedition. He wanted to color in every bit of curiosity and to awaken a new sense of wonder. He wanted everyone to feel welcome. There was warmth and good feeling all around.

My mother was a big part of this. To reiterate, some have suggested that she played a more pivotal role in establishing the tone and tenor of life aboard the *Calypso* than my father—and in this, I must agree. La Bergère reigned over my father and his crew like a venerable queen, taking great pains to ensure that all around were made comfortable and appreciated. She came by this endeavor quite naturally, so perhaps it's inaccurate to suggest there was any "great pain" involved on her part. Her efforts in this regard were, well, effortless. She cast a beguiling spell over the ship and gave the *Calypso* her personality. This showed itself in a thousand small details, all the way down to our cabin assignments. As boys, Philippe and I often found ourselves sharing quarters with some of the younger members of Dad's team. Space on the ship was tight and crowded, and we were often there for only a short visit, so it made no sense to give us our own cabin, but my mother made sure that we bunked with crewmembers who might make a special effort to look out for us. She wanted us to learn from them in ways we could no longer learn from our own parents.

Once, early on, we were sharing a cabin with André Laban, a chemical engineering student who would go on to develop one of the first underwater cameras designed exclusively for my father and who would stay on as Dad's primary cinematographer for more than a quarter century. He first sailed on the *Calypso* in 1952 as a young man of 24, and Philippe and I could only marvel at his many interests. He was a talented painter and musician in addition to his developing skills as a diver. La Bergère, in her own way, encouraged this, as she did with all the members of Dad's crew. If one demonstrated a particular talent, she would coax it out of them, and that appears to have been the case with André Laban.

Every evening, he'd shut himself into his cabin with his precious cello and make the most beautiful music, which he frequently played to soothing effect—like the time I was recovering from a free-diving accident. I'd damaged my eardrum and had been confined to quarters with a violent earache. Despite the many medications and salves my mother administered, I could get no comfort—until Laban entered the cabin one evening after dinner. Without saying a word, he set the cello gently between his knees and drew the bow lovingly across the strings. He commenced to play the sweetest, most beautiful music, and my pain fell away as if by magic. It struck me then, and still, as the most generous gesture for a young, accomplished man to go out of his way to soothe a boy in just this manner. I believe my mother created a certain spirit of generosity that permeated the ship.

Laban himself offered his own take on my mother's role: "She calls the *Calypso* 'my ship.' And, inured against seasickness, she smokes, she drinks, she swears like a true sailor. . . . She also knows, during a stopover in port, how to change herself into a woman of the world, ready to host the French consul or the governor of an island nation. If the need arises, she uses her knowledge of English or Japanese with great success."

La Bergère was a mother not only to Philippe and me—"the two little birds" of Les Roches—but to the entire *Calypso* crew, my father included. And it was largely because of her that I look back on my childhood as an idyllic time. Yes, JYC was the one who captured the world's attention; yes, JYC steeped us in wonder and adventure; but it was my mother who provided the ballast to our unconventional lives. She gave our adventures weight and meaning. Others in my situation might look back on their boarding school years as an anxious point of pause, but I never saw that period in quite this way. With the arrival of the *Calypso* in our lives, there also came a necessary shift or progression in our circumstances, that is all. I fell in the habit straightaway of counting the days until the next school holiday, when I could join my parents to continue our grand adventure. Of course I would have much preferred to have abandoned my formal schooling entirely for the unpredictable adventures that awaited me on the high seas, but that didn't mean I resented my time at Les Roches. I did not feel "unwanted"—as many boarding school students have described. I did not feel trapped, or left behind, in any way.

It was merely my lot to go through these motions. Philippe's, as well. And so we went through them—painstakingly, at times, but we pressed on. Before long, I found a subject or two to excite my imagination in the classroom, and here I attached my father's curiosity and my mother's determination to my own intellectual pursuits. I still wasn't much of a student, but I was becoming more of a thinker. I began to excel in the natural sciences, which I suppose was Dad's influence on full display. His interests had now become mine, and I followed up my years at Les Roches with a short stint at the Janson-de-Sailly College, in Paris, where I prepared for the baccalaureate exam I'd need to take to be admitted to university. Despite myself and my lackluster abilities in the classroom,

I would be the first of the little Cousteau "birds" to take flight, at around the same time my father was severing the last of his ties to the French Navy. He hadn't sailed for the navy in quite some time, but he'd never formally resigned his commission, either—not until 1957, when he was named director of the Oceanographic Museum in Monaco. For the first time in a great many years, Dad had a kind of anchor, in terms of a profession. The appointment was by no means a desk job—to be sure, he would continue to sail with my mother and his team on their *Calypso* expeditions—but it was no formality, either. The position offered a measure of stability, and a steady paycheck that would now be added to his pension as a retired naval officer.

The principality of Monaco was very much in the news at the time, after Prince Albert Rainier III captured worldwide attention with his storybook wedding to the beautiful American actress Grace Kelly. Suddenly, the shores of Monte Carlo and its environs were seen all over the world as a playground for the rich and famous, and it was no surprise that my father managed to soak up some of the spotlight for himself and his work. Certainly, his own burgeoning celebrity might have played a role in getting hired at the museum, where Dad's high profile on the world stage was undoubtedly seen as a great asset. In any case, there was a good and proper fit, and we Cousteaus moved among our Monegasque neighbors as if all eyes were upon us.

Surprisingly, my father began to receive his first bit of negative press during this period, as a few skeptics and pundits weighed in with their thoughts about why Captain Cousteau would choose to live in a tax haven like Monaco. I suppose such speculation was unavoidable, but here it was utterly misplaced. People wondered what my father was hiding. Such nonsense. Dad didn't have any money to hide—in fact, he hardly had any money at all. Over the years, he'd never earned more than a living wage. Whatever income he did manage to draw from his films and expeditions, he immediately invested in his next production or adventure; and, now he was making only a modest salary as a museum employee. So the notion that he was "hiding" anything was preposterous. And yet, there it was.

I took the opportunity to enroll at the College of Monaco so I could be near my parents as my father embarked on this new sideways adventure. I believe my mother welcomed the company. She never said as much, but I could tell she wasn't especially happy to be marooned on dry land for an extended period. From the moment she first set sail on the *Calypso,* she considered it her home, and to her the idea of living any place else was anathema. She'd especially hated the idea of living in Paris, and so Monaco appeared as a more acceptable alternative. However, my parents did not live in the residence normally assigned to the director of the museum. Instead, Simone arranged for an apartment on the top floor of a high-rise building, with a view of the ocean. This alone was no small feat, because the housing market in Monaco was prohibitively expensive, but I'd learned never to underestimate my mother's persuasive charms. Somehow, she found an apartment within my father's means. From her terrace, she could see the harbor where her precious *Calypso* was moored when she was not at sea.

She seemed to call to her, my mother always said.

The Oceanographic Museum was a thriving enterprise. The building itself, where the museum was housed, was visible from my mother's terrace. It was built in 1910, under the auspices of Prince Albert I, who by all accounts had been fairly fascinated with the sea and its vast wonders. In this respect, the man was something of a visionary—and certainly ahead of his time. It's quite an impressive structure, set into a rocky cliff overlooking the harbor, not far from the palace, and conceived in the grand style of the period—like many of the municipal buildings on the Rock, as the locals call the principality. The first time I saw it, I thought it looked like a magnificent cream puff.

The museum's holdings were equally impressive, including exhibitions and collections of various aspects of ocean life, extending at the time to over three thousand species of fish. (Today, the collection encompasses over four thousand species.) A big part of my father's job was to stand as the public face of the museum and its many initiatives, which were operated under the auspices of the Oceanographic Museum of Paris, on which it still depends. As director, Dad enjoyed special stature as a French employee residing in Monaco, although he was among the last to be so privileged, since General de Gaulle abolished the practice in 1959.

The appointment coincided with a noticeable bump in my father's popularity. Already, owing to the worldwide success of *The Silent World* and his various innovations in diving and oceanography, he was something of an icon among the French, but now that he was at the helm of one of the largest oceanographic institutes in the world, he appeared to enjoy a newfound international stardom. The stature of one seemed to fuel the stature of the other, although I cannot say for certain which side benefited most from the association. In any case, my father's books, documentaries, and lectures were becoming wildly popular. And, he especially enjoyed the imprimatur that came with his position, which he was sure to attach to the opening and closing credits of his films, hereinafter produced "with the scientific support of the Oceanographic Museum of Monaco." It proffered a kind of seal of approval on his work.

His "work" at the museum was not without some controversy, although the most controversial incident took place behind the scenes, away from public scrutiny—and at first it was only me who had noticed any sort of problem. You see, Dad and his team were

concerned early on that the museum's gate receipts were not sufficient to fund the many programs and research initiatives underway.
And so what was the solution to increase attendance? Well, reflect-
ing the general views of the period, the team decided to develop a
dolphin exhibit, which they believed would be a popular attraction.
Indeed, many aquariums and museums around the world featured
marine mammals in captivity, so it was inevitable that they would
think along these lines.

How do you develop a dolphin exhibit? Well, you need dolphins and must capture them. So Dad authorized a team to head
out into the Mediterranean, just off the coast, where schools of
dolphins were known to swim and play. These men would literally
scoop up the dolphins with a net as they played by the bow of the
ship. Then they took the dolphins back to an Olympic-size pool
that had been set aside to help them adapt to captivity.

It was altogether an objectionable, onerous process. About
a dozen dolphins were captured in this way, and they were all
placed in this giant pool that was nevertheless overcrowded with
all these glorious creatures swimming about. Even as a teenager, I
could see something wrong with pulling these magnificent animals
from their natural environment, imprisoning them, and crowding them into this unsuitable habitat simply for our amusement.
It was unconscionable, really. And yet it was the order of the day.
No one seemed to think anything was wrong with the practice,
so I didn't say anything. I suppose you could say I was complicit,
because I was actually invited to the pool, along with many others,
to help the dolphins acclimate to their surroundings. I accepted my
assignment joyfully, because I was drawn to these animals. We were
meant to hold them quite still if possible, so that they would slowly
get used to the idea that they were living in a totally new environment, with walls and limitations.

It did not occur to us that we were being cruel, only that we were helping the dolphins make the necessary adjustments. The freedom they had enjoyed in the sea was no more, and this was now all they had, and they struggled against it at first. Very quickly, I began to notice three types of dolphins—at least among our small sample group. There were those who refused to accept their new environment, who resisted all attempts to help them adjust and who quickly died. There were those who gradually adapted to being imprisoned in this contained environment—which really was rather like a jail cell, now that I was forced to consider it—and who could be trained to repeat certain behaviors through a system of reward and punishment. And then there was a third group, caught somewhere in between. Their personalities took a while to adjust to their new surroundings. It appeared as though they were always looking around, considering, trying to make sense of what had just happened to their world as they had known it, and yet, they were able to make certain adjustments and somehow survive.

Among our 10 to 12 dolphins, I could see them line up accordingly, into one group or other. The "professionals" assigned to this acclimation project recognized these distinctions as well, and the fates of these dolphins were soon organized according to their tendencies. Those that died were removed from the pool. A few others that appeared to be struggling, in one way or another, were returned to the Mediterranean. However, a third and final group of "survivors" was transferred to an even smaller pool. The switch was meant to reinforce for these animals that their world was becoming smaller and smaller, where their freedoms would become more and more limited. Of course, we did not have sufficient space at the museum to create a large exhibit with an Olympic-size pool, so it was necessary to get the dolphins comfortable in a confined space to match the one we would ultimately provide.

Of this group, three or four were designated for display. I became quite attached to one of them. I was in high school at the time, and my school was located directly across from the museum, so I fell into the habit of visiting the dolphins every morning. Without fail, I'd arrive in the early morning to feed and play with the dolphins and to help with their adjustment to captivity. Without realizing it, I found that I enjoyed a special connection with these dolphins—a deeper connection than I felt toward any of the course materials we covered in my school curriculum. They seemed to speak to me—or so I had convinced myself. As the son of the museum director, I was able to feed the dolphins myself, and I took full advantage of the special benefit. I considered myself extremely blessed. I had always been attached to every aspect of the sea and its creatures, but here that attachment was made tangible and purposeful. I cast myself as a true caretaker of these fine mammals—and I came to consider one dolphin in particular as "my" dolphin. We had a special relationship. I did not name the dolphin, but I became extremely attached to it. And it was drawn to me as well.

However, after a few days, I began to notice something wrong with this particular dolphin. It was suddenly listless. Where it had once been playful and energetic, it now appeared tired and withdrawn. This went on for another day or more, until I finally called the veterinarian for a medical opinion on the matter. I said, "Doc, there's something wrong with this animal. Can you check it out for me?"

But the doctor could find nothing wrong. He said, "Its heart is okay. Its breathing is okay. The dolphin appears healthy in every way."

Of course, the doctor was looking at the dolphin from a physiological point of view, while I was studying the psychology. I called again a few days later, after I had noticed that the dolphin was becoming ever more lethargic. It was the same scenario. The doctor checked it out and found nothing. I even mentioned the dolphin to my father, but he did not see any cause for concern because the veterinarian had reported that the animal was fine. But it was not fine. I knew it in my bones. It felt to me like anyone could see it—if they would only look! Something was off, most definitely, something terribly wrong, and after another few days, I arrived at the pool to find that my worst fears had materialized. The poor creature was dead. What had happened, we determined, was that the dolphin had swum madly from one end of the pool to the other, as fast as it could. Then it smashed its skull into the concrete wall. This was a suicide, plain and simple. Nobody called it that, of course, but I was sure of it. We had killed that helpless dolphin by our mistreatment, our disregard.

The event was quite a harsh lesson for me, I'll say that. It made me realize that it was unacceptable to put these marine mammals into our little man-made jail, because surely that's what it was. I pressed my father to see the point—and, soon enough, he did. He came to realize that we were not just imprisoning these defenseless creatures, because our cruelty did not end there. It's like we're putting them into these jail cells blindfolded, because their primary sense was auditory, and so we piled one cruelty on top of another. In a limited space like a swimming pool, the acoustics were such that sound reverberated and became almost unbearable for them. Soon, they did not communicate at all. They became completely quiet, sullen, disinterested. And, by some outrageous oddity, no one was crying out against this type of treatment but me, a mere teenager.

Ultimately, the remaining dolphins were returned to their home waters of the Mediterranean. They were never put on display. I was somehow able to convince my father and his colleagues of their recklessness, their irresponsibility, and they set about finding some other way to increase attendance. It would not be at the expense of these animals, after all, and I counted this as an important victory. Small, but important. Even today, all these years later, I cringe when I learn of an aquarium or museum warehousing healthy dolphins for entertainment. They train these animals to jump through hoops or to perform all of these other circus tricks that they do not perform in nature, and it's an abomination. That said, I'm not against aquariums. I can even support the display of certain marine mammals, in certain situations. In Long Beach, California, for example, a blind sea lion is on display that is quite popular with tourists. To return that animal to the sea would be like a death sentence, because it cannot feed itself, so here I have no problem with the exhibit. And so I'm not opposed to taking care of the ones who could not survive on their own, just as I would not be opposed to taking care of a bald eagle with a broken wing. In that case, you take the bird and keep it comfortable, well fed, and perhaps even use it for educational purposes. You bring it to schools, teach children about endangered species, and make the best of a bad situation. But I could not support making a bad situation out of a perfectly good one, merely to boost ticket sales. It's wrong. And, it's an insult to the public. It's an insult to our children, who we bring in for the purpose of educating them about various aspects of nature. But it goes against nature, so it's antithetical to the entire process. It takes us further from our goal.

In any case, my father was impressed by the passion I had shown on behalf of these helpless dolphins. He was heartbroken for me, of course, over the "suicide" of my dolphin friend—and, I suspect, he was a little heartbroken himself, now that he had begun to see our treatment of these animals in a different light. He never said as much, but I believe he started to look on my actions in this matter as an awakening of a kind, a show of maturity that he would now do well to consider, and he took the opportunity soon after the remaining dolphins had been returned to the Mediterranean to press me on my future plans.

"Jean-Mine," he said, when I showed up at the museum one morning, "have you given any thought to what you might want to do with your life?"

In truth, I had not—and yet, in some respects, I had thought about it constantly, in the manner of a child who keeps a careful inventory of his hopes and dreams. Up until this moment, I had thought about my future in terms of my parents' presence, which was to say that I wanted to continue on in the life they had set before me. It was basic. But I had never carved out a specific role for myself in that life beyond wanting to be along for the ride, and it took hearing from my father on this for me to start thinking proactively about making some sort of meaningful contribution. I could not be a passenger on life's journey, I realized; after all, this was my father's life, my father's journey. The *Calypso* was his ship, not mine. I needed to live my own life, and set off on my own journey, so I set about charting my own course.

One thing was certain: I wanted to live and work in connection to the sea. I had no desire to make the navy my career, much to my mother's disappointment. She would have liked nothing more than to see me follow in Dad's footsteps, and in the path set out by her own father and grandfather, but that was not for me. It would not be for Philippe, either. For a while, I gave some thought to studying law. I was helped along in this notion by my mother, who recovered from her disappointment long enough to encourage me to apply my personality in this way. She would tell me that I questioned everything. "Whatever the subject," she frequently said, "you manage to argue the point, like an attorney." She said this so often, I started to believe that perhaps this was the path meant for me, but then I set it aside when I realized I didn't quite have the stomach for law school. I could not see myself working behind a desk, so I looked for another path, to take me where I wanted to go.

More and more, I found myself thinking like my father, assessing man's limitations underwater and seeking new and creative ways to surpass those limitations. It was the approach he took in developing his now-famous Aqua-Lung, and the scuba apparatus that had opened up the world's oceans to recreational divers and scientific research. But even Dad's innovations had their limits, and I wondered how to surpass these as well. This was the way of the world, yes? The thresholds and obstacles of the past are only memories in time. High-speed air travel, telecommunication, space exploration—by the end of the 1950s, man had broken through barriers in each of these areas to achieve what had once been unthinkable. Already, in less than two decades, underwater exploration had undergone a dramatic transformation, and here too, what had once been unthinkable was now at hand. However, like any experienced diver, I realized there was an imperceptible line I could not cross underwater. For every diver, that line is different. Some can dive to impossible, unfathomable depths while coping with the physiological difficulties; others can only plumb closer to the surface; but every diver has a boundary of some kind or other. He can only descend so deep, or remain submerged for so long, before having to return to the surface and breathe fresh air once more.

I was out to change all of that, I started to realize—only it would be some time before I could imagine how.

My father's work in this area had been well chronicled. However, I hadn't been old enough or aware enough of his experiments in diving physiology as he and his friends were conducting them, so I now looked to their findings with great interest. I pumped Dad with questions. He was not the sort of man to find contentment in an initial discovery, so he himself had a great many questions. When he answered one, it invariably introduced another. And then another one after that. I started out wanting to learn everything my father knew about decompression. He wanted to know everything there was to know and then a little bit more besides; such were the differences between us. In time, this once-removed, hand-me-down knowledge would no longer satisfy my curiosity, but for the time being, it was enough. I only wanted to understand the "intoxication of the depths" my father wrote about in his books, and how variables like fatigue, anxiety, lung capacity, and overall physical fitness factored in to an individual's ability to reach certain depths without incident or ill effect. The stages of decompression, Dad reminded me, are entirely predictable, although some divers are more susceptible to them than others, and they occur at different depths depending on all these other variables. But there is no avoiding them.

It was generally accepted that there was no avoiding them, but I wondered if this was truly so. Dad wondered, too. It sometimes seemed that this was the biggest, most consuming aspect of his work—wondering if there was a better, simpler way to a desired end, or if what we'd long recognized as limitations were merely hurdles to be overcome. Together, we wondered if there was some way to dive even deeper or to remain underwater for longer periods. It was like an intellectual puzzle, and we were determined to solve it. Already, my father had considered the development of a private submarine, which he laid out for me and the *Calypso*'s chief diver, Albert Falco, one afternoon over lunch in the wardroom. To illustrate, the captain took two dishes and set them out on the table in front of us so that the edge of one dish was touching the edge of the other. And then, with a flourish that seemed to call for great fanfare or a drum roll, he said, "Here it is, our submersible!"

JYC's latest bold idea was to create an environment big enough for two people, who might be able to slip inside a compartment at normal atmospheric pressure and then remain at normal pressure as the compartment was submerged. It was an elegant solution to a complex problem, and he expanded on it in such a way that the "oceanaut" divers in his small submarine cabin might be able to propel themselves through the water without assistance. It was a clever design, and he followed it up with another invention in support: water propulsion nozzles, or jets, which could be fitted to this personal submarine device and allow the divers to maneuver the vessel to change direction and depth.

Now, the innovations Dad laid out for us on the table were not entirely new, I realized—but that's how it goes with innovations. You take something that already exists and you improve upon it. That's essentially what happened with the Cousteau-Gagnan regulator—and here my father sought to upgrade the helix propellers already in use in some diving vessels, which could often become entangled in cables or nettings underwater. He sought to tweak, to improvise, to refine. The propellers were workable, but they were also problematic, so why not try to eliminate the problems? By great good fortune, he was able to arrange for financing for his experiment through his various contacts in Monaco and in the international film community.

In all his designs and improvements, Dad sought to achieve the most streamlined, most natural experience for the diver. His ideal, always, was that less is more. Les mousquemers and their diving cohorts used to gently mock the "amateur" divers in their midst for sporting so many tubes and gadgets and rough edges to their gear. Such add-on devices could only slow them down and make them less efficient in the water, they always said. The objective, as ever, was to slip through the currents with ease and efficiency. "The less resistance we face," JYC used to say, "the more efficient we are."

To be sure, virtually every one of Dad's innovations was conceived on this principle, and as he further developed this personal submarine concept, he set it up so that tandem divers would each lie flat on their stomachs, each one in his own chamber, each one facing an observation porthole, each one able to operate the steering controls, each one in a position to help or film the other. By the time I was 20 years old, Dad's first "saucer" had given rise to a new line of small submarine-type vessels, including the *Sea Flea,* which was designed for one person. Later, he produced a small sub known as the *DeepStar,* for Westinghouse, and the *Cyana,* built for the French government. Ultimately, Dad's diving saucer and mini-subs were able to operate safely at depths up to 1,250 feet—although the smaller, more efficient *Flea* model could safely descend to 1,840 feet.

And yet, despite these innovations and enhancements to existing technology, despite the way they opened up vast new territories of the ocean for exploration, my father felt something was missing with his "new and improved" diving equipment. And, surely, there was—namely, though all these bells and whistles might allow divers to descend ever deeper toward the ocean floor and to remain at extreme depths for longer and longer periods, they ultimately could only offer an unsatisfactory experience. It was a revelation he might have had sooner, before all these governments and technology companies invested so much time and talent on this enterprise, but my father eventually came to the conclusion that humans needed to touch, feel, and manipulate objects with their hands in an ocean environment. A tactile sensation, he believed, was necessary to a pure and rewarding experience of underwater exploration—and that sensation was just out of reach to a diver cocooned in one of Dad's submersible shells. His initial impulse, back when he was merely trying to keep up with his free-diving friend Didi, had been to swim like a fish, to experience the ocean's flora and fauna firsthand. He'd set out to become a manfish, and he and his friends had succeeded in this beyond their wildest dreams. But now it appeared that each innovative step forward had only taken him a few steps back, farther and farther from the tactile experience he'd longed for as a young man.

Make no mistake; these submersibles had their necessary applications. They would prove to be essential tools for filming and study. These first primitive models led to more advanced designs that oceanographers and underwater scientists are still using. But there would be no joy in putting them to use. No fascination, no wonder—at least, not for a man like my father. They fascinated him as he saw his concepts and theories being realized, but they could not hold his interest. They would be our tools of the trade, nothing more. This was a disappointment, to be sure, but only a small one. I had been as excited about these developments as anyone, but I could certainly appreciate my father's point. He might have succeeded in the manner of solving a theoretical exercise, but in the end he had not succeeded in shattering the limitations of divers all over the world; he had merely traded one set of limitations for a whole new set of limitations, and in the trade-off, he was no closer to his long-held goal.

Yet my father could not curb his desire to descend ever farther, ever deeper, for longer and longer periods. It was in his nature to break through these barriers, even if they were barriers he himself had established. It was inevitable, then, that he eventually turned his thinking toward the development of underwater habitats. This concept, too, was not entirely original to my father, but when he put his mind to a prospect, it helped to bring that prospect closer to reality—and here I put my mind to it as well. In fact, I put my mind to it most of all, because before too long, as I followed my father to all of his meetings and listened to his ideas, we were deep into the design of what could only be termed a "house," which he hoped to place at the bottom of the sea and outfit in such a way that it might be suitable for a diver to remain there for an extended period.

Such an exciting prospect! Such possibilities!

To this end, JYC began working with engineers at the French Office for Underwater Research, which in French translated to the acronym OFRS. (Soon, the office would be known as the Center for Advanced Marine Studies, or CEMA.) The collaboration continued for several years, as I completed my degree in my chosen field of marine architecture—a field that did not exist as such, I should mention. By 1961, JYC and his team commenced our so-called Continental Shelf, or Conshelf experiment, to determine if our design and concept had any practical applications on the ocean floor.

I threw in with Dad and the Cousteau team on a kind of grand plan. I had already decided that I would light out on my own at the first professional opportunity, but I was in no great hurry to seek out such an opportunity. I could still learn a great many things at my father's knee, a great many discoveries we might pursue together. He could be wildly enthusiastic about an innovation or development, in much the same way he had torn open that shipping box from Aire Liquide to try out the first prototype of his regulator. And, as anyone who ever met the captain could attest, his wild enthusiasm was infectious. Here, he got to thinking that man could somehow colonize the continental shelf—the landmass that reached from the continents below the surface of the sea. He considered the depths between zero and 1,500 feet to be the richest and most interesting part of the ocean for study. I agreed with him on this, although I thought we might be getting a little bit ahead of ourselves in terms of colonization.

Still, we called our experiment Conshelf I—or Précontinent I—with the dream that others might follow. Dad arranged the necessary financing, and after several months of final preparations, we placed our completed "house" at a depth of 33 feet on a rocky shelf just off the coast of the smallest island in the Frioul Archipelago, about a mile from Marseille. The island, known as If, is dominated by a castle made famous by the author Alexandre Dumas, who used it as a setting in his novel *The Count of Monte Cristo.* Looking back, I can only smile at my father's choice of location for this experiment, because the name of the island itself suggested hope and wonder and possibility.

It begged the question of the ages: What if . . . ?

For the first Conshelf inhabitant, Dad tapped one of his favorite, most dedicated divers on the *Calypso* team—Albert Falco, whom we all called Bébert, a native of Marseille who was so passionately and desperately crazy about the sea that he never once set foot on shore during his career on the *Calypso.* Bébert would make the descent with his diving companion, Claude Wesly. The men believed it was a great honor to be tapped for this assignment, although their duty did not promise any sort of luxury accommodations. Indeed, the "house" could have been more properly described as a steel barrel. It was painted yellow and measured 16.5 feet long and 8 feet in diameter, with an entry shaft at the bottom. The living conditions were surely spartan, at best. Already, CEMA specialists had tested the long-term impact of residing at such depths on animals such as sheep, goats, and pigs—to no ill effect—so the men weren't especially worried about their safety. (Bébert joked that if a sheep had been up to the task, then he could handle it as well.) Rather, they worried about the psychological effects of being contained for seven days in such a confined space, but their excitement seemed to outpace their worry.

In his excellent account of the experiment and his career at my father's side, *Capitaine de la* Calypso, Falco described the mood inside the underwater "house," which he and Wesly had dubbed Diogenes, after the Greek philosopher who had famously renounced all but the barest necessities to pursue a life of the mind. "The sixth day is marvelous," he wrote. "I enjoy my status of oceanaut. Never before had I taken the time to look around me at the bottom of the ocean. There is no need to be obsessed by a chronometer! I see every piece of sea grass, in great detail. Some of the plants have flowered— pretty, discreet corollas, in a shade of yellow-green I have never observed before. At night, in the light of our projectors, the underwater prairie teems with life. Sea anemones and ceria unfold their panache of tentacles. . . . Crabs stride along the seabed. . . . Some starfish are red, others ocher. . . . Sea horses hold on to each other by their tails. . . . Claude and I begin to forget the world above us. . . ."

Falco's enthusiasm was also infectious, and just a few weeks after the conclusion of our Conshelf I installation, which I briefly visited as a guest diver, I made a decision. I would be an architect, I determined—but not just any architect. I would be a marine architect, and hope to someday build houses beneath the ocean. I would do this on my own, just as my father had pursued his dreams on his own.

Now, here is an interesting contradiction about my father, a man who had built his life and reputation as an impossible dreamer: He was also something of a realist, especially when it came to the career choices of his children. In his mind, it was one thing to pursue an abstract experiment such as the one we had just undertaken off the shores of the Frioul island but quite another to devote all of your resources toward the pursuit of a far-off ideal. And so he took a pragmatic view. He said, "Jean-Mine, do you really want to build houses under the sea? Are you quite sure?"

"Sure and certain," I replied.

He smiled indulgently, as if he might have been lecturing a small child—which he probably thought he was. He said, "You must give this some more careful consideration. You must realize your opportunities will always be limited. You must realize there will always be less construction beneath the ocean than. . . ."

Here I interrupted him—perhaps for the first time in my life. "I know," I said. "There will always be less construction beneath the ocean than on land, I'm not a fool."

"No?" JYC shot back. "Perhaps not, but surely this is folly." He dismissed my enthusiasm for this endeavor as an extension of Bébert's enthusiasm for his seven-day tour.

Perhaps it was, but I would not be dissuaded. Even La Bergère could not talk me out of my determination. To my youthful, indulgent thinking, it did not even matter that there was no school of underwater architecture where I might pursue my "folly"—if indeed it was just that. In fact, there is still no such school. However, I decided to get a conventional degree in architecture and then see what I could make of it. I would pursue my dream in what ways I could, even if it was a pipe dream.

Shortly after graduation from the Paris School of Architecture in 1964, I had the great good fortune to be put in contact with a Parisian architect who was keenly interested in tubular construction, and he encouraged me to visit the naval dockyards at Saint-Nazaire for further training. There, I would be better able to grasp the distinctions between conventional architecture and naval architecture and to consider the technical constraints facing engineers in a hoped-for future that might include underwater architecture. In any case, I could consider myself a marine architect at long last, and I couldn't have been happier.

Well, I suppose I could have been happier. It would have been nice to have a client or two, or at least a firm prospect, and therefore the respect and admiration of my parents, but this was a small quibble to an inveterate dreamer cut in the Cousteau mold.

Regrettably, Dad wasn't overly optimistic about my professional future, even as he continued with his own experiments in underwater living. This time, his plans were far more ambitious. There had been tremendous, worldwide excitement following the Conshelf I trial, so much so that he now looked to film a documentary of his follow-up experiment. For Conshelf II, instead of just one "house," he assembled an entire "village"—with a main structure known as the Starfish, for the way it appeared to boast three tentacles emerging from a primary structure. This time, Dad chose a location in the western reaches of the Red Sea on a reef off the coast of Sudan. It was a hopelessly beautiful spot—selected, in no small part, for the brilliant, clear-blue color of the water and the panorama of coral that would surely enthrall filmgoers. Among his many gifts, Dad was a true showman, and he knew that for audiences to respond to his planned documentary, they would need to be transported by the splendors of the sea.

Once again, Albert Falco was asked to lead Dad's underwater team, along with Claude Wesly, a cook, and two marine biologists. Also, Wesly had a stowaway—his pet parrot. He had tried to bring the bird with him for the seven-day experiment on Conshelf I, but my father refused to allow it. Here, though, for a 30-day stay, JYC relented. He could not see the harm, he finally said. Moreover, he probably recognized the family-friendly, public relations value of having a parrot on board. It would be good for the film, he probably thought. And besides, miners had quite legendarily taken canaries with them into coal mines. If the birds perished due to lack of oxygen, the miners knew to return to the surface as quickly as possible or else they'd suffer the same fate, so here it was possible for onlookers to regard Wesly's parrot as a kind of air pressure alert system, even though it was more pet than harbinger.

In addition to the main sleeping compartments for the five Starfish inhabitants, there were also several hospitality chambers, where guests could stay for short visits. There was a separate submarine-type hangar for the diving saucer. A second two-man housing chamber was stationed at a greater depth—82 feet—and at some midpoint in the experiment, two additional divers would remain there for as long as a week to further test the limits of human endurance in these conditions.

The five Starfish oceanauts were visited daily by a team of researchers and scientists, who would record their physiological changes and mood swings. My father also made several descents, to check on his team over the course of their month-long experiment. The main chamber was anchored at a depth of 33 feet, and it was really quite comfortable—or so I was told. There was food and wine, music and television. There was even air-conditioning. And, most incredible, the men were all able to smoke their precious cigarettes throughout their confinement, because they were breathing fresh air. And so, whenever my father returned to the surface after one of his visits, he would announce that conditions below were quite lovely. "It's like they're on vacation," he once joked.

Even La Bergère descended to Starfish for a visit, thus becoming the world's first female oceanaut. She was meant to join my father for only one night, to celebrate their wedding anniversary, but she was so enamored by her underwater environs that she stayed on for another few nights. My parents toasted each other with flat champagne—because the bubbles, of course, could not announce themselves beneath two atmospheres of pressure.

Today, nearly 50 years after Dad's Conshelf II experiment, I look at some of the reality shows featured on American television, such as *The Real World* on MTV or *Big Brother* on CBS, and recognize that my father was something of a pioneer in this genre as well. You see, he and his producers had outfitted their Starfish set with cameras at every imaginable angle. The idea was to chronicle the movements of his crew and to generate footage of their most mundane interactions. The resulting documentary, *World Without Sun,* was enthusiastically received when it was released in December 1964. It earned my father an Academy Award for Best Documentary. What audiences found especially compelling, we learned, was not the unique window JYC opened on his underwater world so much as the chance to observe ordinary human behavior in extraordinary conditions. It was a fascinating human experiment that recorded the subtle and not-so-subtle changes taking place in Dad's "village" inhabitants. Some of these changes were physiological, as a result of the changes in pressure, and some were behavioral, as a result of the long confinement. For example, the personalities of the oceanauts were altered dramatically and so were their tastes. After only three or four days, they were no longer interested in reading the letters they received from friends and family, although they couldn't get enough of them at the beginning of their stay. Soon, they stopped responding to them. (I suppose if there had been such a thing as e-mail back then, we might have noticed the same disinterest in keeping connected.) Each day, they became more and more detached from the world above the surface and more and more focused on the interactions taking place in their confined space below the surface. Even their pastime pursuits began to change. They started out listening to popular music, but very quickly selected only classical pieces. They started out reading popular adventure and detective stories, but abandoned these for more serious material. They started out dining on hearty, heavy French fare, but before long began asking the cook to prepare salads and grilled dishes.

Most interesting of all were the changes in their temperaments. The good humor that characterized their personalities on descent was difficult to find after only a few days. The men, who had been easygoing and adaptable at the outset, were soon complaining and grousing over every little thing. And yet their good humor remained at the surface of their personalities. For example, in the deeper cabin, during the week-long experiment that called for the divers to breathe a mixture of oxygen and helium, the men could not remain angry at one another for the length of a single argument. Anyone who has ever inhaled the contents of a helium balloon and immediately tried to speak can understand why: The men sounded like Donald Duck. Surely, it's difficult to mount a serious argument or vent a personal frustration when you sound like Donald Duck. And so the divers had only to open their mouths to speak before being reduced to laughter and good cheer once more.

My studies prevented me from traveling to the Red Sea for the Conshelf II experiment, but I looked on with a great rooting interest. Here, my father's experiment appeared to dovetail directly with what I had recently decided would be my life's work, and so his findings would form an essential baseline. And yet I was reluctant to consider the trial a great success. Yes, the men had all survived the ordeal with no ill effect: Some indicated they could have happily remained below the surface for an additional stretch of weeks; others were quite happy to be done with their confinement. And yes, Dad's cameras were able to record a thrilling documentary that would excite filmgoers around the world. But, to a man, all involved agreed that it was unreasonable to believe that humans might someday "live" under the ocean. We might outfit ourselves in such a way that we could comfortably survive there for extended periods, as Dad and his team had just demonstrated, but for the next while at least, the idea of actually living there would remain the stuff of science fiction.

For me, it was a disappointment tucked away inside a discovery. Over time, I would come to recognize the consequence of Dad's pioneering experiments in underwater living, but at just that moment, I thought I might need to rethink my career plans. However, what we learned from Conshelf II—and, in 1965, from a Conshelf III experiment conducted in the Mediterranean near the Cap Ferrat lighthouse between Nice and Monaco—was that men (and women) could live and work in an underwater environment for sustained periods. This alone was a thrilling discovery. For Conshelf III, six divers, including my brother Philippe, resided at a depth of 328 feet for three weeks, during which time they successfully worked at various industrial simulation tasks, including the construction of an underwater oil well.

And so the takeaway for me, son of the great Captain Cousteau about to light out on my own in a theoretical discipline that at the time had no real practical applications, was that my father had once again succeeded mightily in opening a window onto another world.

It was up to us to pass through that window and see where it might lead.

8

From Madagascar to Hollywood

For a time, I wondered what I might do with that architectural degree in my pocket, but a plan of action presented itself soon enough. In this, I had some help from the French government. You see, France was at the tail end of the Algerian War, and I was of military age so I had no choice but to serve. I enrolled as a noncommissioned officer, hoping to get through my required tour without incident.

By great serendipity, I was sent to Toulon for training—specifically to the barracks in Mourillon, the district where I was born. It was not a happy homecoming. I hated everything we were taught there. I hated having to use a gun. I hated the idea of confronting an enemy. I hated the ways we were made to interact with one another and with our superiors. I even hated the condescending manner in which we were taught, but I stuck it out. However, when I had completed my training, destiny played a joke on me: The war in Algeria was over. I would not have to fight, after all.

This was, indeed, a fortuitous turn, but I still had to fulfill my obligations so I made a slight detour and enrolled in France's version of the Peace Corps, known as Coopération. Once again, I had to endure another interminable period of training, during which I was actually asked by one of my supervisors if I had any specific expertise that might prove useful in the field. I answered that I had studied architecture. "Great," came the reply, "then you can peel potatoes."

Such was the organizational efficiency of a French bureaucracy.

But as I had learned observing my father's Conshelf experiments, it is possible for man to withstand almost anything—and here I allowed myself the small carrot of dreaming about my deployment. I had put in for assignment to Madagascar, and sure enough, it came through. I was overjoyed. I had long dreamed of visiting this island nation off the southeastern coast of Africa. It seemed so exotic, so untouched, so rich with history. For the next two years, I lived and worked in the capital city of Antananarivo, tooling around in a dusty, old, blue Renault my father had given to me some years earlier. I was 27 years old, and most of my work involved the design and construction-supervision of six schools to be built in some of the more impoverished, more remote coastal regions of the country. We were also charged with building small residential units—the local equivalent of low-income housing.

The assignment offered a peculiar lesson in the vagaries of human behavior. Here we had thought we were making significant improvements in the lives of the local people we intended to serve, but our efforts were not always appreciated. For example, we fitted our modest houses with indoor plumbing, offering what we hoped would be received as a major improvement in hygiene and comfort. But we

As my father is getting ready for a dive, two crew members are already suited up. (Robert B. Goodman, NGS)

My father always knew how to get his point across.
(Thomas J. Abercrombie, NGS)

The Calypso *in her heyday. (Luis Marden, NGS)*

As director, my father explains a planned expansion of the
Oceanographic Institute in Monaco. (Gilbert M. Grosvenor, NGS)

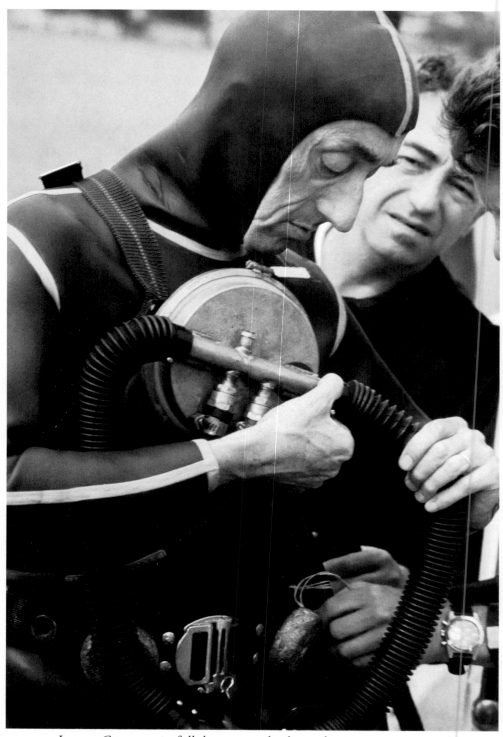

Jacques Cousteau, in full diving gear, hooks up the Aqua-Lung, assisted by his deputy and captain Jean Alinat. (Bates Littlehales, NGS)

My father directs the launching of the diving saucer in a series of test descents off Puerto Rico. The two-man saucer could hover anywhere within its 1,000-foot depth range. (Thomas J. Abercrombie, NGS)

The Conshelf III, the third underwater habitat, is being launched at the harbor of Monaco, to test the survival skills of six oceanauts for 27 days at 300 feet. (Bates Littlehales, NGS)

My father in his element, filming a sea lion on Lobos Island, Uruguay.
(James P. Blair, NGS)

Standing discreetly beside my father is my mother, Simone. Affectionately called La Bergère (the sheep keeper), she looked out for everyone on the team. (©Tim Trabon)

Here I am conferring with my father, diver Bertrand Sion, and Paul Martin, Calypso engineer, from our river raft, the Pirarucu, during the Calypso's Amazon expedition in 1982. (©Tim Trabon)

*President John F. Kennedy awards my father the National Geographic
Society's Special Gold Medal in the presence of Simone Cousteau.
(J. Baylor Roberts, NGS)*

*The next generation: My daughter, Céline, and my son, Fabien, with me during a
film shoot in the Dry Tortugas.(© Carrie Vonderhaar, Ocean Futures Society/KQED)*

Standing discreetly beside my father is my mother, Simone.
Affectionately called La Bergère (the sheep keeper), she looked
out for everyone on the team. (©Tim Trabon)

Here I am conferring with my father, diver Bertrand Sion, and Paul
Martin, Calypso engineer, from our river raft, the Pirarucu, during the
Calypso's Amazon expedition in 1982. (©Tim Trabon)

President John F. Kennedy awards my father the National Geographic Society's Special Gold Medal in the presence of Simone Cousteau. (J. Baylor Roberts, NGS)

The next generation: My daughter, Céline, and my son, Fabien, with me during a film shoot in the Dry Tortugas. (© Carrie Vonderhaar, Ocean Futures Society/KQED)

did not take into account the habits of the Malagasy, who preferred to do their "business" in the great outdoors. The idea of putting one of our bathrooms to its intended use was an abomination to these people—and, as such, an essential reminder to take into full consideration the customs and habits of the people I hoped to serve.

It's an elemental insight, wouldn't you agree? The moment we understand and accept the way other people live—other species, even—is the moment we can find a satisfactory solution to whatever problem we are confronting. My father understood this concept especially well. It was among his many gifts to be able to step outside himself and see a situation from another point of view. He saw the world as a sailor, as a businessman, as a moviegoer, as a fish. And the world looked back admiringly, for he couldn't help but understand. He placed himself in the perspective of others, and in this way found the points of connection that made his work so memorable, so meaningful.

Over time, I endeavored to do the same.

I drove often along the dreadful roads of Madagascar, behind the wheel of my secondhand Renault, and I came to know the island better than many of its inhabitants—who for the most part had no access to vehicles of their own. Like many other islanders, they were drawn to the sea. Even those who lived several miles inland seemed to make their way to the water; it was the lifeblood of the region. Life in Madagascar can be harsh, but the setting is undeniably beautiful. In the south, the coast is lined with beaches covered with fine sand and surrounded by lush coconut plantations. To the north, the seascape is cut by coral reefs that reach from the shore to a small island outpost known by the pretty name of Nossi Be. Of particular note, I thought at the time, was the way most of the Malagasy in my acquaintance were blissfully unaware of the contours of their island. In general terms, they were a warm, intelligent, sensitive, hardworking people, and yet their indifference toward maritime matters had undoubtedly left them exposed and vulnerable throughout their history and continued into the present, as other nations looked to capitalize on their natural bounty.

I thought of those early lessons I had taken with regard to the disinterest in indoor plumbing and came away thinking there was no way to help these people until and unless they were inclined to help themselves. And the only way to get them to that point, I realized, was through education, which was why building those schools was such a rewarding project.

In some ways, I could have stayed on in Madagascar indefinitely—although in others, a two-year stay was time enough. I felt at home on the island, never more so than when I was diving along the coast, getting to know the waters of the Indian Ocean, its seabed, its resources. I never suspected that I would have occasion to return to these shores in any sort of professional capacity—but, alas, I fear I am once again getting ahead of my story.

The time away from my family had a positive side effect, because in many ways my brother Philippe began to fill some of the spaces made by my departure. He was much more of a showman than I had ever been, much more like my father in his gregariousness and charm, and he gradually took a front-and-center role as "Cousteau's son" during this period. The cameras "loved" him, as the saying goes, and he returned the affection. I set this out with no jealousy or rancor. Philippe's emergence in this way was a good and welcome turn, because I was quite happy with my sideline role in the family "business." I had made a special attempt to separate myself from the ongoing scientific and charter operations of the *Calypso,* and I hadn't participated at all in Dad's Conshelf III experiment. I wanted to swim in my own waters, I suppose you could say. Philippe, too, was very much his own person, but his talents as a filmmaker, editor, and diver cast him in my father's pool, so the overlap worked well for both of them. He was quite content, swimming in the same waters as my father—and quite adept at it. More and more, he took on an active role in my father's professional life, as that life became his as well.

There was a telling moment that took place during the Conshelf III experiment, when Philippe was cocooned with five other divers for nearly a month. At some point, a journalist visited the control room above the site and asked my mother if she would submit to an interview. He asked, "Madame Cousteau, aren't you worried about your son?"

In response, my mother flashed a stern, reproachful look—as was typical of her—and said, "Sir, I'm not worried about my son. I have six sons down there, not just one."

She was, after all, La Bergère—the ship keeper, the sheep keeper, and the keeper of our fates and fortunes. It did not much matter to Simone Cousteau that one son had set off on his own adventure, while another served as a human guinea pig for one of JYC's grand experiments. We were all children of the *Calypso* as far as she was concerned, and this applied whether we were on board, or diving from her deck, or pursuing a separate dream on a distant shore. The men kept watch over the ship and over the sea, while she kept watch over us.

Meanwhile, Dad's career continued apace. The success of *World Without Sun* was welcome and unexpected, matching and perhaps even surpassing the first splash he'd made with *The Silent World*. However, studio investors believed it was too soon for him to follow up his account of the Conshelf II experiment with another full-scale documentary of his efforts at Conshelf III, so he turned to television. Or, to put a fine point on things, television turned to him. In his own way, my father anticipated this, for just as I was learning to study the habits and customs of the local population of Madagascar, so, too, was Dad putting the same lesson to good use in the United States. In the early 1960s, American television offered a dynamic pipeline to the American people. The power of the medium in the United States was far greater than it was in Europe—or anywhere else in the world, for that matter. Dad recognized this, I believe, and he shot extensive footage at Conshelf III and flew to Washington to share it with producers for the National Geographic Society, which had been a great and early supporter of his work. Soon, the footage had been edited down to a one-hour special, which aired on CBS in early 1966, with narration provided by actor and director Orson Welles. Once again, the production was met with great enthusiasm and positive review attention—and with a magnetic voice of authority like Orson Welles to help tell the story, it's no wonder audiences were rapt.

My father told me of the pending CBS deal upon my return from Madagascar, and I couldn't have been happier for him. It was the culmination of his lifelong dream to be able to share his films with so many people, all at once. Yes, he had already enjoyed fantastic success with the releases of *The Silent World* and *World Without Sun*, as well as dozens of shorter documentaries, but what excited him about this American television deal was the chance to reach as many as 20 or 30 million people in a single showing. In contrast, a hit film might be seen by only a few hundred thousand people, a best-selling book might only sell in the tens of thousands, so the difference in reach and impact was potentially enormous. He was plainly thrilled—and, as ever, his enthusiasm was infectious.

The world could not contain Dad's many interests and talents as a child, an adolescent, a student, a naval officer, a diver, an oceanographer, a poet, a pianist, a painter. He was always looking to do something more, something else, something different—and yet, above all, he wanted to make films. This had been the one and abiding constant in his life, reaching all the way back to the 9-millimeter Pathé camera he acquired as a boy. He wanted desperately to share his unique vision with audiences all over the world, to tell stories people could not experience anywhere else, to let others see and experience what he was privileged to see and experience. And in this, he had certainly succeeded.

Now he was poised to succeed on an even grander scale—in a smaller, more intimate medium, to be sure, but a grander scale just the same.

The man in charge of turning all of that Conshelf III material into one hour of riveting television was a young producer named David Wolper—another fortuitous turn, because I believe it's fair to say that David Wolper had more to do with Dad's success in American television than anyone else in Hollywood. Wolper liked to tell the story of how he watched Dad's completed National Geographic special and realized that the fish on the small television screen seemed to come across like fish in an aquarium. Color television was a relatively new phenomenon at the time, and most American households were still watching television on black-and-white models, but Wolper remembers this realization as a kind of lightbulb-over-the-head moment. He'd never met my father during the editing of the CBS special, but now the brash young producer thought he should get to know "that little Frenchie" and determine if there might be a future for him in American television. He thought television audiences would respond to a series of specials featuring Captain Cousteau and his *Calypso* team and their various undersea adventures, so he flew to Monaco to sell my father on the notion.

It wasn't exactly a hard sell, I'll say that. I wasn't there for the meeting, but I was told that Dad received Wolper like he had been waiting for him for some time. A part of my father probably wondered what had taken these guys so long, but he was happy for the attention. In fact, he was thrilled at the prospect, not least because it promised to solve some of his perennial money problems. Well, perhaps *problems* is too strong a word in this context, because Dad always managed to earn a living. He and my mother lived quite comfortably, but nevertheless modestly, and traveled the world as if on a whim; but there never seemed to be enough money to fund his various projects and expeditions. Perhaps it would be more accurate to call them money worries, for surely he worried how to bankroll his adventures. Here it appeared he might be able to pursue his interests in a more full-throttled manner, with backers to underwrite his research and exploration in such a way that he might surely move about on a whim.

Of the three main television networks, only ABC expressed serious interest in an ongoing series based on Dad's underwater adventures. CBS had broadcast Dad's most recent special, to great reviews, but network program executives didn't think a regular series would be profitable, and they believed it was too soon to air another one-shot special. The ABC offer did not begin to cover Wolper's estimated preproduction costs, which included a complete makeover for the *Calypso* and newer, more television-friendly diving suits and dive tanks for the Cousteau team. In all, Wolper calculated that JYC would need to spend nearly $1 million to refashion the ship and its crew for their television debut. Dad didn't have that kind of money—and neither did Wolper. However, the producer turned to sponsors such as DuPont and Encyclopaedia Britannica to make up the difference and in the end, delivered a deal for 12 hour-long episodes. At first, ABC was only interested in airing four one-hour specials, but Wolper could be a rather convincing fellow.

Looking back, these details all seemed to fall into place fairly quickly, although at the time, my father felt like the negotiations were dragging on interminably. Nevertheless, Dad was overjoyed at the opportunity to bring his undersea world to American television audiences—and to collect whatever fame might come his way in return. Understand that fame would be exponentially greater than any he had known to this point; after all, it was one thing to move about in the limelight conferred by a best-selling book, a hit documentary, or an Oscar or two, but quite another to be anointed a star of American television. Frankly, I don't think my father had any idea what was about to hit him. He was not prepared to become a celebrity. On one level, he understood the power and reach of the medium, but on another, he probably thought he'd still be able to move about in relative obscurity. And the money! It was as much as he could have hoped to earn in several lifetimes, just for this first dozen episodes. According to some accounts, he stood to earn more than $4 million from the initial ABC order—a veritable fortune in those days.

By this point, my brother Philippe was working full-time at my father's side. In Hollywood terms, he was like a supporting actor. As I have written, he was quite comfortable in front of the camera. As the footage from *World Without Sun* and the Conshelf III National Geographic special confirmed, he was a showman in the Cousteau mold. He was also adept at negotiating many of the behind-the-scenes details necessary for a successful underwater expedition—arranging for permits, contracts, shipping, supplies, and so forth. The two men made quite a team, and together with my mother the *Calypso* was well staffed with the Cousteau touch.

I was content with my sideline role, although I was only too happy to check back in with an assist from time to time, as the situation warranted. This happened with some frequency, once production began on the television show, which the network had decided to call *The Undersea World of Jacques Cousteau*. Dad was certainly pleased with the title, but mostly he was pleased with the platform on which he now had to share his perspective on the world's oceans with American television audiences.

One of the first examples of my father calling on me to help was also one of the most illustrative. It was early on in the production of the first 12 episodes, months before the first show even aired. The idea for the series was to focus each hour on a specific aspect of marine life—giant sea turtles, shipwrecks, dolphins, and on and on. (Our oceans offered such a vast canvas; the producers would never run out of subjects.) Later, Wolper would hire Rod Serling of *Twilight Zone* fame to provide the narration, helping to lend Dad's series a distinctive yet familiar sound. Typically, Dad would start each episode with a question or a common myth about the ocean, and then he and his divers would go looking for answers or ways to shatter that myth. There would also be some of the "real world" elements that had made Dad's documentaries so successful, and so the cameras would once again be turned on the *Calypso* and her crew. Here, as before, the production would succeed on the personalities of the Cousteau team, which now numbered approximately 25, plus a giant Saint-Hubert hound given to my parents by Princess Grace of Monaco as a kind of "going-away" present before Dad's first televised voyage. My mother named the dog Zoom, for reasons that were never made entirely clear to me at the time, although she later explained that she was feeling like they were all about to set off on a thrilling new ride to explore the great unknown, and she thought the name was evocative.

The first episode focused on sharks, with new footage to be collected primarily in the islands of the Maldives, in the Indian Ocean off the southern tip of India. However, the *Calypso* experienced engine trouble before she ever got there, so for a while, it was thought they might double back and hope that existing footage on sharks they had recorded in the Red Sea might sufficiently supplement the new material they had just recorded for the first episode. They also thought they had enough material for a second episode on coral reefs, should they decide to reroute the ship for repairs. However, JYC and his crew ultimately decided to forge ahead on one engine and hope for the best—reaching the northern Maldives a few days behind schedule, although there was time enough to capture the breathtaking footage they needed to launch the show in trademark Cousteau style. From there, they were off to Kenya, where Dad knew of a shipyard capable of handling the repairs.

Just as the *Calypso* became seaworthy once more, and they were sailing into the Red Sea, fighting broke out in the Middle East, in the famous Six Day War between Israel and Egypt. Dad was caught in the crossfire, I'm afraid. It was June 1967, and the ship was briefly a target of the Israeli Air Force. Gunners began firing upon her early one morning in a glancing attack that lasted but a few minutes before Israeli officials realized their mistake, although it was time enough to scare the living daylights out of the ship's crew as they slept below decks. Mercifully, she did not suffer too much damage—only cosmetic, which I suspect caused our new friend Wolper a good deal of distress because the *Calypso* herself was now a television star, sporting an expensive facelift. It would not do to present her to audiences with bullet holes and pockmarks in her hull.

But the *Calypso*'s troubles ran deeper than appearances—and beyond the Israeli military, as well, because the Egyptians kept boarding the ship to inspect its contents and to monitor the crew. A tense, uncertain stalemate followed that would extend well beyond the six days of the conflict itself. The entire region seemed on lockdown, amid a flurry of bombs and bullets. No one knew what might happen next—even, what was happening still. The Suez Canal had been closed, and the *Calypso* was prevented from returning to her home waters of the Mediterranean. Communication was difficult—or, impossible. It would be months before the canal would be accessible again, so Dad and his team had to scramble. They had a production schedule to keep, which meant that for the first time in Dad's career as an adventurer, there was a clock—or, at least, a calendar—to consider.

First, it was determined that Dad would leave the ship as soon as possible, while Philippe remained on board with my mother and the rest of the team, awaiting further developments. Next, the crew made plans to sail the ship around the Cape of Good Hope, which meant a longer voyage at greater cost. The rerouting also meant the ship would need to be resupplied, but the Egyptians prohibited this for a few days. They also kept Philippe and Didi from bringing supplies to shore, such as Dad's precious film, which now needed to be processed and prepared for editing. The impasse reminded me of the stories JYC used to tell about carting his film canisters across France on his bicycle during World War II for processing and safekeeping.

Once Dad managed to land in Djibouti on the Horn of Africa, he reached out to me in Paris, where I was living with my wife, Anne-Marie, and our young family, making my first tentative forays into the world of French architecture. He knew I was familiar with the region and thought I could be of some help. It was not to be the first time he sought my assistance, and it would not be the last. Each time, I dropped whatever I was doing and answered the call. I did this, I suppose, because I felt a kind of siren's pull to the sea, to the *Calypso,* to my family. Also, as I have written, my father could be very persuasive, especially with regard to me and my brother. We were so eager to please Dad that he could get us to agree to almost anything.

The ABC program was an important opportunity not just for my father but for my entire family, so I told myself that my brand-new profession—and my family—could wait. My father needed me. My mother needed me. My brother and the rest of my extended *Calypso* family needed me as well. Of course I would answer the call. This merely meant seeking to make the best of an unforeseen situation. As long as the *Calypso* had to sail around the Cape of Good Hope, we might as well take advantage of the opportunity to shoot some captivating, never-before-seen footage that Dad might be able to use for the show.

We would multitask, before the phrase was even coined.

My first thought, knowing he was obliged to sail around Africa, was to make a stop in Madagascar, because I knew the area so well. There would be much to explore in those waters, I knew, so I set about making preparations while the *Calypso* continued to wait out the aftermath of the Six Day conflict. When she was finally ready to set sail, I had all the authorizations in place for her passage to the Indian Ocean. All the red tape had been set aside. There would also be stops along the way in the Seychelles and on Europa Island in the Mozambique Channel. My father even laid out an ambitious plan to sail across the Atlantic, perhaps continuing through the Panama Canal as far as Peru, where Dad had always wanted to film in the waters of Lake Titicaca and search for the legendary treasure of Inca gold. Without the backing from the television show, such an expedition would have been unthinkable—but it was an exciting time in the extended Cousteau family as we all awaited the show's debut.

It was on Europa Island, in fact, that we came upon a gathering of green turtle hatchlings—thousands of them, it turned out. We had read about them, but we were quite unprepared for what we saw. Our cameras were fortunate to capture the turtles' movements as they literally ran toward the sea. However, a great many of these hatchlings never made it to their destination, because a flock of hungry frigate birds was soaring above, and they would swoop down and collect the tiny turtles in their beaks. One by one, they'd pick them off, like ducks in a shooting gallery. Of the select few that did make it all the way to the sea, many would be met near the shore by hungry fish that would also eat their fill.

And so we were treated to a magnificent display of natural selection, played out to full dramatic effect for our cameras. It was thrilling and disturbing, all at once, because of all the hatchlings that set out for the sea, only a very few made it. And life continued on as before.

Some months later, Dad edited the dramatic footage of the hatchling turtles for his second installment of the ABC series, entitled "The Savage World of the Coral Jungle," which aired to extremely positive reviews and strong ratings, and reinforced the feeling among our team that it was a time when anything seemed possible.

And yet, like any extended sea voyage, we could only proceed one leg at a time, and it fell to me to ensure that our first legs would yield some worthwhile footage. Happily, our next fortuitous turn in this regard found us in the waters off the Iles Glorieuses— a grouping of tiny French islands in the northern Mozambique Channel, between Madagascar and Africa. Later on, Dad's cameras captured some wonderful footage of sea lions romping in the water—such delightful, playful creatures, with their splendid coats of shiny, brown fur. Dad took a special liking to two of the sea lions in particular, which he named Pepito and Cristobal. He thought they were incredibly smart, and he was fascinated—and smitten—by their behavior. He especially admired their outgoing personalities and the ways they interacted with the divers, and on another one of his whims, he decided to "invite" them to sail with the *Calypso* on the next leg of her journey, after I obtained a permit from the South African government.

Over the years, Captain Cousteau would receive his second round of criticism over this decision, but at the time, he reasoned that "capturing" Pepito and Cristobal would allow him to study their ability to adapt to new environments. In the context of the times, I suppose such treatment of these creatures was fairly benign, although today conservationists and animal activists would look upon it as an act of cruelty, to remove these sea lions from their natural habitat primarily for the entertainment of American television audiences. Even worse, the cameras recorded hours upon hours of Dad and the crew treating Pepito and Cristobal like circus animals. There is no excuse for this, and yet in 1967 we were not as attuned to the fragility of animals and sea creatures. The conservation and preservation mind-sets of today had not yet taken hold, so researchers and scientists of my father's generation did not always give our environment the respect it deserved—again, no excuse, but perhaps a helpful explanation.

Even at the Oceanographic Museum of Monaco, where Prince Albert I's compassion for the marine world was unquestioned and unrivaled, early studies of dolphins often involved harpooning them and dissecting them—practices that would be unthinkable today. I offer this observation as a kind of framework, because even the "pasha" himself disavowed his own treatment of Pepito and Cristobal in later years. At the time, though, no one saw the harm in bringing them on board; to the men, they were only a welcome distraction.

When they are in relative infancy, sea lions are far more trusting and agreeable than they are in maturity. That explains how these two creatures were so easily "captured" in our nets as they lay basking in the sun with their group on an outcropping of rocks. However, once on board the *Calypso,* they remained docile for several days. They hardly moved. Perhaps their malaise was psychological, JYC reasoned. They'd never known anything but freedom, so it made perfect sense that they resisted their confinement in our cages designed to protect divers from sharks. They made their displeasure known, in what ways they could.

After a few days, they appeared to relax. They stopped baring their teeth when one of the men approached with fish in hand. All day long, day after day, they were plied with fresh fish and squid, and they came to look forward to their feedings with the good nature of a contented gourmand. (I believe the cook prepared some other treats for them as well.) Soon, they were allowed to move about the deck of the ship without constraint, and the cameras recorded many playful moments when the sea lions appeared happy and content. At one point, after crossing the Atlantic to Tobago Bay, our divers sealed a small cove with elaborate netting, and the team returned Pepito and Cristobal to the water to play with them and study their behavior. They were used to the cold water of the Indian Ocean, but they adjusted to the warmth of the Caribbean as if they had been born to it. After some hours, their separate personalities began to emerge, in a way the men hadn't seen on the ship. Pepito emerged as the more high-strung of the two; Cristobal was much more relaxed, more trusting, perhaps even faithful. Both were quite responsive to repetitive forms of play.

The team spent several hours each day swimming with Pepito and Cristobal in the harbor, and after a few days, Dad determined that the sea lions had acclimated well and could be trusted to respond to the call of their human friends. With this in mind, he ordered that the netting be removed, so that the sea lions might swim more freely—but this only proved to be a bad idea on top of Dad's initial bad idea, for there followed a misadventure of embarrassing proportions. The "faithful" Cristobal immediately made for the open sea and disappeared along the horizon. JYC and his men were so stunned by his departure that, for a moment, they made no move to go after him, although it's unclear to me just how they might have kept pace with a sea lion if they had thought to try. Instead, they were merely dumbfounded and crestfallen as Cristobal raced out to sea.

It was such a sad, dispiriting outcome—but Cristobal's story did not end there. By some odd chance, he was recaptured a few days later, about 62 miles from the harbor, by a Puerto Rican fisherman who promptly sold the poor creature to a misguided American woman who thought it might amuse her wealthy friends if she had a sea lion in her swimming pool. It was difficult to imagine a more transparent abuse, although to be fair, her actions were only slightly more misguided than my father's in capturing the animal in the first place. But the moment Dad read in the local newspaper of Cristobal's whereabouts, he set off in pursuit. His charms were on full display, and he somehow located the woman and cajoled her into releasing the sea lion into his custody—for a much higher price than she had paid to the fisherman, I might add.

It was, in all, a costly experiment—with a sad ending. Dad had intended to demonstrate to viewers of his new television show that man and sea lion could coexist quite happily and establish a genuine friendship. This turned out to be the case, to a degree, but it was not a friendship of the sea lions' choosing. They did not ask to be removed from their herd, from their familiar waters, from their natural habitat, or from their routine. They merely adapted, and found a way to get along—to make the best of a new circumstance. They learned that if they behaved a certain way, they would be treated to a handful of squid. But they were not meant to live aboard an old wooden minesweeper. Ultimately, the sweet and trusting Cristobal developed food poisoning after eating a buffalo toad and died soon after. The men were all devastated—and so was Pepito, who appeared lost without his playmate. So Dad made arrangements to release him along the Peruvian coast, in an area well populated with colonies of sea lions.

And yet, even this final kindness was ill conceived, because certainly our Pepito stood a much better chance of surviving and thriving in the Indian Ocean than in these waters half a world away. Yes, he could once again swim with his own and move about freely and naturally, but Pepito's "cousins" in this part of the world were entirely different from the ones he had left behind. Their reconnoitering and communication signals were different. Their feeding habits were different. Their environment was different. And so we never knew whether our friend was able to adapt to these new waters and to these new companions that were so like him and yet so different.

We could only hope.

Over the years, Dad would look back wistfully at this odd interlude with the sea lions. He could not be wrong, even when he was; it was against his nature, and it was impossible to see this as a strength or a weakness. It was merely an aspect of character. It was who he was. Privately, my father even expressed remorse at the mishandling of these creatures. But publicly, he defended his actions to the end. He said, "Our objective was to demonstrate that marine animals are capable of almost as much affection as land animals, and we achieved that objective."

Well, I suppose he did just that, but by the time *The Undersea World of Jacques Cousteau* debuted on ABC television in January 1968, he was already on to his next adventure.

TREASURE

We Cousteaus were scattered across the globe as *The Undersea World of Jacques Cousteau* approached its January 1968 premiere, and the long wait between planning, filming, postproduction, and broadcast was unsettling for my father. To be clear, it was a happy-anxious time for all of us, but for JYC most of all. He was used to being in more direct control of his work, and it was difficult for Dad to grasp that while he and his team were swimming with giant turtles off the coast of Africa, executives at ABC were busily preparing to make him a household name. The one seemed to have little to do with the other, even though it was all part of a piece.

It was also difficult to imagine what the ABC series might mean for my father and his team going forward. For many years, Dad had experienced a certain degree of

177

fame in France and to a lesser degree throughout Europe. Beginning with the publication of *The Silent World,* he moved about in a kind of spotlight. It wasn't an especially bright spotlight, but those who followed his advancements and accomplishments certainly appreciated his contributions in underwater exploration and oceanography. However, now that appreciation was about to reach a whole other level. That spotlight would now shine brighter than ever. Already, JYC had enjoyed a taste of what was to come with a promotional tour in New York, during which reporters asked him every conceivable question about his life and work and the program he was about to present, including why he thought television audiences might respond to a program about sharks. Dad responded as though he couldn't understand the question, but his confusion had nothing to do with a language barrier. His English was fine. Rather, he couldn't imagine that a person wouldn't be interested in such dramatic footage of such savage creatures. It was like asking him why he thought people enjoyed breathing air. To him, nothing was more entrancing than watching these sharks up close, even if that meant through a camera lens and a television screen. For most Americans, that would surely be up close enough.

I was in South America when the program debuted, scouting locations and seeing to arrangements for the Cousteau team. This had become virtually a full-time job, causing me to set aside my plans to put my architectural training to good and productive use, at least for the next while. Philippe, who had no such plans to separate his work from Dad's, was nevertheless at his new home in California with his wife, enjoying a brief moment of pause. La Bergère was on board the *Calypso,* off the coast of Africa, and Dad was returning to her from his brief junket in New York.

Wherever we were, it didn't much matter, because our world would never quite be the same; the program was an immediate and resounding success. The glare of the spotlight was brighter than any of us could have imagined, and as we endeavored to fulfill the rest of our first-season order, we began to experience the impact of having a hit show on American television. The impact was subtle at the outset, especially at sea, but it found us soon enough. Before it could, though, we had to prepare for the next expedition, and now that Dad had pulled me back into his swirl, it meant that most of these preparations fell to me. I was like a field producer—or, in political terms, an advance man on the campaign trail. By necessity, I was always one step, one country, one adventure ahead of the team.

These days, it's possible to make such arrangements in haste, but communication was slow and unpredictable back then, so I often had to make exploratory visits into various ports of call to determine our needs and help our team negotiate any difficulties. For example, at around the time the television show was premiering in the United States, I was scrambling to make arrangements for the transport of two small submarines, which I hoped to ship by train to Lake Titicaca, at an altitude of 12,650 feet. Can you imagine? Submarines on a train! Headed up a mountain pass! No one on the receiving end of my phone calls or telegrams had heard of Captain Cousteau just yet. No one could quite understand our peculiar shipping needs. I suspect these negotiations might have been easier if the other parties knew of my father and his work, because as it was, they had no choice but to think I was crazy. A few months later, they might have known that Dad was the crazy one, not me.

Another example: I was waiting for the *Calypso* to arrive in Lima, Peru, when a coup d'état forced President Fernando Belaunde Terry to seek exile in Argentina. I did not know him personally, but I had long admired the man from afar, not least because he was also an architect—yet another who was not presently working in the field, I should add. There were tanks up and down the streets beneath my hotel window. There was commotion and excitement all around. It was a dangerous scene. I was a prisoner in my room, but my concern was not for my own safety as much as for the *Calypso*'s. I worked desperately, frantically to get word to the ship to avoid the region.

"Whatever you do, don't come," I cabled.

It was a one-way distress call; I never received a response.

I called the French embassy several times each day for an update on the situation, without success. It could have been a disaster for our team if the ship arrived in port amid such turmoil. Mercifully, the revolution did not last long— until I stumbled on it—and three or four days, at most—but they were the longest three or four days of my life to that point, filled with dozens upon dozens of frantic telephone calls and dispatches attempting to explain who we were and what we were trying to accomplish.

By the time the *Calypso* arrived, a semblance of order had been restored, so we continued on toward Lake Titicaca as if nothing had happened. However, there had been a sea change we couldn't help but notice; the local government was in relative disarray, so it was naïve to expect a return to normalcy—and, sure enough, things were far from normal. At each stop along the way, we presented our travel documents, which had been signed by authorities no longer in charge; in many cases, these individuals had been exiled or imprisoned, and here we were trying to gain passage on their now-invalid signatures. There were a great many tense moments, as functionaries considered these documents, which were invariably deemed official, and in this way, we managed to proceed without incident.

Our stated mission was that we were off in search of Inca gold, which had long been a dream of my father's, ever since he'd read about the storied waters of Lake Titicaca as a boy. And now here he was, a soon-to-be-famous world explorer with a generous subsidy from ABC television, DuPont, and Encyclopaedia Britannica, so he believed the time was right to make such an exotic, complicated voyage. And complicated it certainly was. We traveled by train to Puno, a small city on the shores of Titicaca, in the southeast corner of Peru, and we must have made a confounding picture; there were 17 of us in all, laden with diving tanks, compressors, diving suits, extensive camera gear, and those two small submarines—altogether more than 20 tons of equipment. The train followed a magnificent route, but it was also steep and treacherous. At several points, the train was so burdened by the climb and the pressing weight of our equipment that we had to stop and back up to negotiate the terrain.

The sheer beauty of the countryside overshadowed the danger and the difficulty of the trek. The scenery was thrilling and quite unlike any landscape I had seen. We were all enthralled. To pass the time, we counted 56 different types of trees as we climbed that mountain. Llamas and alpacas were grazing outside our windows the entire way—resilient, all-purpose animals for which the Peruvians seem to find a use in every aspect of their lives: for their milk, their meat, their hides, and their stubborn strength as pack animals.

Our arrival astounded the locals. They could not believe their eyes when we stepped off that train and began unloading all that equipment, even though I had been to the area some months earlier and had warned them what to expect—a warning that could not have prepared them in the least, I now realized. What an astonishment for these people who lived under simple circumstances, primarily in floating villages built of rushes and logs. The Uros, the oldest inhabitants of the region, were known as "the people of the rush," and it was said that they had arrived on these shores well before the Inca. For centuries, the Uros had lived above these grand waters, but they lived in fear of them at the same time. It was an unusual contradiction: They were superstitious and believed they would die if they fell into the water, but that didn't stop them from eating the fish; it just kept them from swimming.

The Uros could not understand why a group of men would dive into the lake with such brio, such enthusiasm. Or, why we would dive into the lake at all. They had no frame of reference for our behavior. We explained that we had arrived to help them search for their lost treasure. This was a welcome bulletin, because the legend of the Inca gold had been part of their storytelling for centuries. They said prayers asking for its return, and they stood in shocked silence as our team set off on our Zodiac inflatables toward the Isla del Sol—the island of the sun. Shocked silence, it appeared, was their way of cheering us on.

One old man who was said to be more than one hundred years old (and who somehow appeared even older) approached me as we put into the icy cold waters of the lake and told me that there used to be a long, ornate gold chain linking the Isla del Sol and the Isla de la Luna—the island of the moon. He said this with such conviction it was impossible to tell if he knew this to be a legend or if he believed it fully. After all, the islands were nearly eight miles apart. Surely, a necklace of gold stretching across such a distance would be impossible—and, impossibly beautiful to behold. We heard many such stories, as we explored Lake Titicaca—an immense body of water, spanning 3,200 square miles and reaching to depths of more than 932 feet. Most people had never even heard of it before Dad put it on the map for them, except a few who might have liked and remembered the singsong name, and yet it was the highest navigable lake in the world, a truly magnificent natural wonder.

According to the legend we sought to dispel or confirm, the treasure had been thrown into the lake during a ceremonial Inca sacrifice many centuries ago. The locals shared this story with us, asked us to join them in song and prayer, and expressed their hope that we could find the treasure for them—because their innate fear of the water made it impossible for the Uros to search for it themselves. The Uros blessed our tiny submarines in a very touching ceremony led by an honorary virgin. Like all virgins, this one was said to have the ability to perform miracles, although as it happened, she had some trouble performing this one in particular. Even the blessing of our "fleet" yielded only frustration. The people threw flowers at our submarines and pasted candy on the hulls, but, still, we could find no treasure.

Be assured, Dad wasn't motivated by any monetary gain that may or may not have attached itself to any possible discovery. He was simply motivated by the thrill of the chase, the great adventure. The discovery itself—if it came—was the end game, for him. He was an explorer, and here was something to explore. That is all. The lost treasure of Inca gold was the holy grail of underwater exploration, as far as he was concerned. Those of us on my father's diving team were motivated in the same way, although I must confess a not-so-secret desire to discover the treasure of legend. To a man, we were all small boys once more, dreaming of jewels and gold and riches beyond our imaginations. We made several descents into the crystal blue waters of Titicaca, for several consecutive days; we were reluctant to give up on it, but the lake was too deep, too big, too cold, too unfamiliar for us to make a completely thorough search. We gave it our best effort, and yet all we found were some old pieces of pottery—interesting, on the face of it, but not exactly what we were looking for.

After some time, disappointed, we decided to abandon our search, although by this point, we had made an unexpected discovery that would prove to be a treasure of a kind—frogs, thousands upon thousands of them. Perhaps a million or more. You see, inadvertently, we'd come across a strange, rare genus of frogs known as *Telmatobius culeus,* some of them quite distinctive-looking, only not in the most pleasant of ways. They captured our attention, my father's most of all. These unusual creatures can live at depths of up to 394 feet without ever having to come up for air, because their skin is able to absorb sufficient oxygen from the water.

As we made this unexpected discovery, Dad came to an important decision: We were there to make a film, after all. We were there to make discoveries. Our stated purpose was to help the viewers explore and understand the wonders of the deep, to open up a remote corner of the world for their education and enjoyment, and there was no denying that these ugly, unusual specimens were wondrous indeed. And so we turned our attention away from the legendary treasure to these amphibians instead. They moved about as if in a swarm, and we recorded those movements as best we could; the resulting footage was shown in an episode entitled "Legend of Lake Titicaca," which originally aired in April 1969, and introduced the world to these odd little frogs.

In the end, the mass of unusual frogs offered an invaluable lesson—namely, that you might spend your days reaching for some elusive thing or other, but you must always be prepared to embrace whatever you manage to grasp. It was a lesson my father might have articulated to me and my brother many times as we were growing up, but it took experiencing it in just this way for the message to truly resonate. This was JYC's way. This was his approach to life, on full display. No, these ugly frogs were not quite the treasure we'd been seeking, but they ended up yielding an enormous benefit, both for my father's new television show and for the local people. Audiences responded to the creatures in a positive way, and before long, a veritable industry had developed around these frogs. Leading universities and research institutions looked to learn more about this species, which could only be found in plenty in these waters. It opened up a profitable market and a substantial new source of income for the Uros, who soon began harvesting and shipping specimen frogs to laboratories all over the world, where they could be dissected and studied. In this way, I guess you could say we had helped these people to found the *Telmatobius* industry and to put these peculiar amphibians in the spotlight—a spotlight Dad was only too happy to share.

Surely, it was far less glamorous, far less exciting to make a film about frogs than it would have been to record the discovery of ancient treasure, but it was certainly interesting. And entertaining. And ultimately rewarding in its own way. I look back on this adventure and consider my father's role and character. Whatever he did, he was determined to get results. Here, the intended result was just out of reach, so he reached for something else. He was disappointed about missing out on a treasure he had thought about for so many years, but he knew some other outcome would make the long journey worthwhile. Indeed, the value of the journey was in the journey itself, and he recognized this. It was not in my father's nature to worry over matters outside his control. That is why the marketing of his expeditions never seemed to hold his interest. Discovery, adventure, exploration—these were his goals. He did not rest on his laurels or dwell on his disappointments; he would simply seek the next discovery, the next adventure, the next exploration and continue toward it until he found whatever he wanted.

Another man might have looked on our expedition to Lake Titicaca as a failure, but not my father. No, there was no glory for the crew, but we managed to discover important resources that the Uros and Inca continue to exploit for the good of the region.

As an aside, I should mention here that we had the assistance of the Bolivian Navy on our Titicaca adventure. I never fully understood their presence, but there they were. I suppose it was something for these men to do—a productive use of their time. As I understood it at the time, the Bolivian Navy was rather large, especially considering that Bolivia has no seaport. As you can see from a map of the region, it only has one, hardly significant shoreline. It borders a part of the great lake, in the center of which an imaginary line has been established, separating Bolivia from Peru.

The navy boasted a splendid launch, which the United States had donated. One day the captain sent word that we were being invited to a reception in our honor. Dad happily accepted on behalf of the entire team. The *Calypso* crew was always ready for a fine party, and this promised to be a welcome respite from our diving and filming. And it surely was. Our Bolivian hosts were enormously proud of their beautiful vessel; the captain even made a special effort to show it off, ordering the engines to be run at full speed, so he could move the ship forward—and backward—and forward, again. This turned out to be an ill-conceived display, because as he gunned the engines, a propeller slipped off its shaft and sank to the bottom of the lake. It was an embarrassing moment, but a potentially devastating one as well, because without the propeller, the magnificent launch would be reduced to a floating barge.

At this, the captain looked at my father despairingly, unwilling at first to ask for our assistance in the propeller's recovery, but knowing that our divers could probably retrieve it with ease. Dad read the look in the man's eyes and immediately sent a few of us into the lake—and in this way, the men of the *Calypso* managed to save the Bolivian Navy, at least from ridicule.

Another side note: Dad was ultimately proven right to go seeking this elusive treasure of Inca legend. More than 35 years after Captain Cousteau's memorable visit to the region, an international research team announced the discovery of ancient artifacts at the bottom of Lake Titicaca. It was a remarkable find that included stone and metal artifacts and ancient ruins that were surprisingly well preserved at a depth of approximately 230 feet. Researchers traced the discovery to an ancient Tiahuanacu tribe that predated the Inca Empire.

Was this the Inca gold of local legend? Who can say? But Dad and our team made our own treasure from our disappointment.

PHILIPPE

I t would be some time before my father returned the
Calypso to her home waters—more than four years, as it
turned out. Dating from February 1967, when she set sail
from Monaco on an odyssey that would take her to all corners of the globe, the Calypso crossed more than 150,000
nautical miles, in and out of harm's way, and above and
beyond any realistic gauge of her stealth and stamina that
might have been attached to Dad's best assessment when he
found her more than 25 years earlier.

Along the way, JYC's cameras captured more than two
million feet of film and enough viewers to rival some of the
biggest stars of American television. The captain himself
became a celebrity of the first magnitude, as recognizable
to American audiences as Lucille Ball or Walt Disney. Even

my brother Philippe became something of a heartthrob, sharing screen time with my father and emerging as a television star in his own right.

And yet, despite the runaway success of *The Undersea World of Jacques Cousteau* and the attendant rush of fame that found each of us in turn, Dad knew better than anyone that the true stars of the show were the sharks and whales, dolphins and turtles, and penguins and frogs that appeared each week in American living rooms in brilliant Technicolor in such an intimate way that viewers were made to feel a part of the accompanying adventure. At least, that is what he believed, going in. What he hadn't counted on—what none of us had counted on, really—was that this deep personal connection Dad was making with his audience would somehow eclipse the fleeting adventures his cameras were sharing. At bottom, Dad was a storyteller, and together with David Wolper and his talented postproduction team, he spun fascinating tales of the briny deep, each with a kind of beginning, middle, and end to rival any scripted escapade. The human element was a constant, underlying aspect of each episode, but it was never meant to be out in front with those beautiful fish. To JYC's thinking, we humans were hardly more than background players—and, as such, a convenient repository for the abiding interest and affection of television audiences. We came into their homes on a regular basis, in an up-close-and-personal sort of way, and they in turn got to see us live and work in our own home and to survive any number of near misses or close calls. So upon reflection, it was only natural that there was a deepening of interest in my father and his team.

By the end of its long run, 36 episodes of *The Undersea World of Jacques Cousteau* were produced, many of them with Philippe in a principal costarring role, both in front of the cameras and behind the scenes. This footage cemented Dad's reputation in the minds and affections of most Americans, despite the many successful adventures that preceded it and the fine and substantive work that followed. As a result, people remember this period most of all. That's the reach and power of a successful television show, on full display. It can frame your life and work in an indelible way, and so when people today are reminded of the *Calypso* or of my father's pioneering work, their thoughts invariably turn to the footage that emerged from a productive eight-year period. They remember us as we were back then; they turn to the interpersonal relationships that developed on board the ship and the occasional thrills and dangers that found us at sea, as if they all happened yesterday—and they look back at us as if we are much the same.

Alas, that is not the case.

Simone, my dear mother, was a constant presence on these shows as well, but mostly in a behind-the-scenes sort of way aboard the *Calypso*. As an abiding and enduring "character" of these television shows, she was without peer. She was the ship's maypole, at the center of all activity; everything that happened did so at her behest, all the way down to the manner in which the cook prepared our meals.

Me, I stayed involved in the early going in such a way that people remember my role even today, but I drifted away soon enough. I was determined to "do my own thing," to borrow an expression from the period. In the beginning, though, I joined Dad and the crew as we moved from one adventure to the next, often without letup. From Lake Titicaca, for example, the *Calypso* headed straight for the rocky coastline of Guadalupe Island, Mexico, a little more than 120 miles from the shores of Baja California. The trip marked a homecoming of sorts, because it was the first time since the beginning of the Six Day War that I would set foot on American soil—a place I would soon call home.

It was to be a jumping-off point for Dad and his team, resulting in several more memorable episodes before ultimately returning the *Calypso* to her base in the Mediterranean.

For one, JYC had arranged to film a group of scientists who had designed a new type of submersible off the coast of Catalina Island, California, for an episode he planned to call "Those Incredible Diving Machines." And they were surely that—incredible! But something happened to get in the way of Dad's filming schedule. By a bit of serendipity, the *Calypso* was descended upon by a massive school of squid. There seemed to be millions of them; they were beyond counting. And they were everywhere, even in the water intake valves running from the ship's generators, which shut down all power for a time. It was quite a scene. And here again, the forces of nature and natural selection were on full display. The squid would lay their eggs on the ocean floor, and then die or swim away.

Dad believed he had no choice but to momentarily scrap his plans of filming the scientists and their submersibles. They could wait, he determined. The squid, however, could not, and so he redirected his cameras to record this unexpected display, which ultimately became one of his most popular episodes: "The Night of the Squid." For 36 hours straight, he filmed this swarm of squids, including the crew cleaning, cooking, and eating the squid. We had them for breakfast, lunch, and dinner. They were everywhere!

As it happened, so was I. You see, I did not complete the *Calypso*'s journey with her; instead, I set off on a journey of my own. Just as the *Calypso* had been at sea for an extended period, so, too, was I at sea, with respect to my own career. The metaphor was not lost on me, as I moved uncertainly from assignment to assignment, at once eager and honorbound to please my father and help to carry out his mission, but anxious for a time when I could return to the life I had imagined for myself. Oh, I had done more than just imagine it; I had prepared for it as well, and yet in some respects, I was no closer to the life of an architect than when I had set off in my studies. For years, I had worked at my father's side and at his pleasure, and now I had a wife and small child to consider, with no clear picture of what our life as a family might look like once I allowed myself to pursue my own course.

That picture emerged soon enough. Dad sent me to Los Angeles from Lake Titicaca with a couple dozen film canisters to be processed and edited. It was precious cargo. The plan was for me to meet Alan Landsburg, a director my father had worked with on a number of occasions. Alan and I would go on to work together ourselves, eventually capturing an Emmy for a wonderful film we made about the Mississippi River. Our first encounter was memorable. Alan met me at the airport, in a Corvette convertible. I had never driven a convertible sports car, and it was exhilarating. Everything about California seemed fantastic to me, and Alan was a spirited guide. He took me to the Hotel Bel-Air, where I indulged myself in a luxurious bubble bath in a room that was typically reserved for my father. I must admit, it was a welcome extravagance after so many months at sea, with only the bare-bones comforts of the *Calypso* and an occasional cold shower to wash away the concerns of the day. My indulgence followed me to the hotel's fine restaurant, where I enjoyed a sumptuous meal that, together with the elegant surroundings, confirmed that the freezing waters of Lake Titicaca were a world away. I ordered up a fine feast and a bottle of the best champagne, knowing I would charge the bill to my father as a business expense and that this would stand as meager compensation for the work I'd been doing at his side.

From time to time, whenever my postproduction schedule allowed, I joined the *Calypso* for a few days, as she made her way up the West Coast, eventually sailing to Alaska before heading back to the Mediterranean. Along the way, we filmed a giant, 100-pound octopus off the coast of Washington State for an episode called "Octopus, Octopus." And, we visited a particularly headstrong walrus on an ice floe in the waters off Alaska for a program called "The Smile of the Walrus." My poor father spent so much time trying to record the sounds that emanated from this one walrus that it got to where the walrus itself seemed to become exasperated, so he slipped beneath the ice and swam away.

"He's tired of Cousteau!" my father shouted after him.

Despite my frequent comings and goings, I no longer felt quite at home on the *Calypso*, and it would be some time before I could feel truly at home in Los Angeles, but I kept at it. Over the next days and weeks, I toured the city, which struck me just then as an endless strip of look-alike houses lining streets that appeared to run in every direction. There was no "heart" to Los Angeles, I came to realize, no center, which was the opposite of our cities in France and throughout Europe. As a result, I moved about feeling somewhat rootless and disconnected, although I must confess the weather was certainly appealing, and the abundant luxury had some appeal as well. Still, there was much to experience and admire. Opportunities started to present themselves, such as working in collaboration with talented individuals like Alan Landsburg and making important contacts in the fields of architecture and design. Plus, it didn't hurt that the Cousteau name now meant so much more than it had just a couple years earlier. Here, again, I was the "son of Cousteau," and the distinction appeared to mean even more in Los Angeles at the end of the 1960s than it had in Toulon in the south of France at the end of the war.

Without realizing it, I began to feel at home in California and started thinking about putting down roots there. The loose plan had been for our family to return to France, but the feel and pace of Los Angeles was beginning to grow on me; I could see myself and my family there for the next while. I had no money, but the once-removed fame that came with Dad's success on television was proving enormously useful in opening up several doors and opportunities. However, Dad returned before I had a chance to consider any of these doors and opened up one of his own.

He said, "Jean-Michel, I have a job for you."

My father did not ask if I would consider a new assignment. Neither did he ask if I was even in a position to set aside work in my chosen field for work in his. He simply announced that he had a job for me, in such a way that it was clear I was expected to consider it; and, I simply responded in the manner of any adult child of a passionate, irrepressible force of nature like my father. I could not contain my curiosity, but even more than that, I could not turn down his quite reasonable requests. After all, JYC was my father. If he had something in mind for me, I knew I would do well to consider it. And, more to the point, he knew I would do well to consider it. Plus, between the lines of Dad's invitation, there was also the suggestion that he needed me and that my old family needed me—and, frankly, it was nice to be needed.

He came to California one afternoon to meet with David Wolper and members of his postproduction team, so we carved out some time for lunch. He would not tell me ahead of time what he had in mind, only that he needed to talk to me about some urgent matter. Always, with my father, there was some urgent matter or other, a certain element of heat and haste and urgency that came with all of his endeavors, which I suppose was one of the qualities that attracted so many good and adventurous people to work at his side.

Finally, he sat across from me and said, "You are an architect, yes?"

Right away, his approach made me wary. Whatever he was about to ask me, it would be at some remove from the previous role I'd been filling as the *Calypso*'s "advance man." JYC wasn't one to make small talk or meaningless conversation. He knew full well that I was an architect, and he must have also known that I was probably anxious to get started on my career. Yet, for some reason, it must have served his purpose to introduce my training into the discussion. But what was his purpose?

"You already know the answer to that, JYC," I said. "So why don't you tell me what you're really asking?"

It was then that he told me he'd been asked to help with the refurbishment of the R.M.S. *Queen Mary*, the legendary ocean liner that had just been retired from the Cunard line. Ambitious plans were under way for the ship to be refashioned as an exhibition space and tourist attraction, to be permanently berthed in Long Beach, California. As part of those plans, Dad was in consultation to develop and design a theater to present his films in an unusual and appropriate setting.

"You are an architect," he said, this time not asking. "You trained in the shipyards at Saint-Nazaire. You have the necessary qualifications to complete this project. You know my work as well as anyone. You are a diver. Your English is okay, so you will be able to communicate my ideas to your team."

He was right, as always. I was uniquely qualified to represent my father on this project and to lead the design team. In many ways, it was an ideal opportunity. However, JYC had neglected to cover one essential aspect of the job—namely, my salary. For whatever reason, money was hardly discussed when it came to working with my father. At least, it was never discussed with me. When it came to me and Philippe signing on as part of the *Calypso*'s crew, there was a tacit agreement that we would be taken care of, but I never received any sort of steady paycheck or direct compensation. As far as I ever knew, Philippe's role was undertaken on similar terms. Our expenses were always covered, and Dad never questioned me when I put in for an elaborate meal like the one I had at the Hotel Bel-Air, but we were meant to work for the love of the adventure, for the good of the family, for the legacy we were building as ambassadors of the sea. And here I was, at a stage in my life when I needed a little something more than a precious legacy if I wanted to feed, clothe, and shelter my young family.

I said, "But JYC, I have to work. I have to earn a salary."

He said, "I'm well aware, Jean-Mine." (Another sign I should be wary: He only called me Jean-Mine when he wanted to remind me of my consigned role in our parent-child relationship. Simone, in contrast, used my childhood nickname as a marker of affection or a flash point of wistful nostalgia.) Then he flashed his most irresistible smile and said, "That's why I'm offering you $700 a week. That will be your base pay."

The amount struck me just then as an incredible salary—more than $36,000 per year, a very respectable wage at the time. To be clear, the money wasn't coming from my father's pocket. I would be paid by the city of Long Beach. This was the amount that had been budgeted for the ship's redesign—or, at least, the amount my father was willing to allocate to me.

Dad didn't even give me time to respond in the affirmative before sweetening his offer. "If everything goes well," he said, "after ten months, there will be a bonus payment of $50,000, and it shall be yours."

I thought, *Fifty thousand dollars! A princely sum!* And just about what I would need to afford one of the houses I'd been looking at in the area. And so I started work immediately, although to this day, I'm still waiting to receive my bonus. That was often how it went with my father. Ambitious plans and grand assurances were his stock-in-trade. With JYC, the promise was always greater than the reality, but I never resented him for falling short or losing interest when another set of ambitious plans diverted his attention and enthusiasm for whatever undertaking I'd just been assigned. There's no denying that I was well paid, in terms of experience and adventure—just not in the ways I had been promised, nor in the ways I would need to be fully independent.

Soon, the *Queen Mary* project expanded in my mind from a simple theater to a more elaborate museum space. Dad enthusiastically approved my designs and concepts, and a project that was once meant to take only ten months to complete stretched on for three years and more. Perhaps that explains why I never received my $50,000 bonus payment on "completion," because we could never quite find the finish line. Already, there was something else to occupy my father's front-burner attention, and so whenever it appeared we were done with one aspect of the remodeling of the *Queen Mary,* another aspect emerged to consider. The ship finally opened its doors to tourists in May 1971; however, we were still under partial construction, so we were only able to offer a limited array of our planned services, only on weekends. Some months later, in December, we opened Dad's much-anticipated Museum of the Living Sea exhibit—but to only mixed results, I'm afraid. The redesign required a thorough knowledge of ship design and construction, which I had learned in my time at the shipyards. If we had just proceeded according to the grand design Dad produced or to the original plan, the ship would have certainly collapsed under the waterside pressure. The finished space was quite impressive, and initial interest in our exhibit was high; but over the long term, the confined quarters proved unsuitable for many of the exhibits, requiring large segments of Dad's "collection" to be moved to another space, but we never considered the project a failure.

That was one of the curious aspects of JYC's personality. There was no such thing as a failure to him. At least, there was no such thing as a failure of his own making. He had no patience—and no attention span—for that sort of thinking. Once he was through with his end of a project, once it was jump-started and well on its way, he was done with it and on to the next thing. That was how he approached his expeditions, his inventions, his films. Each one brought him a step closer to the next one, and one step farther from whatever he had been working on previously. He would not concern himself with any day-to-day details or ongoing concerns that necessarily followed in his wake. That was how he was in regard to the *Queen Mary* project. He was excited at the outset, curious at our progress, and proud at our opening, but his interest waned somewhere along the way. Ten years later, when most of the exhibits finally closed due to flagging ticket sales and our difficulties sustaining these man-made habitats, he might have even expressed some small surprise that we had been open and operating during all that time.

This was the mind-set he took to the founding of the Cousteau Society for the Protection of Ocean Life, which he began to talk about in the early 1970s. I have long suspected that this was the something else that had claimed some of his focus and passion from the *Queen Mary* project. (Even to a focused and passionate man like Jacques-Yves Cousteau, there was only so much of the stuff to go around.) Dad's idea, he always said, was to develop a member-supported, not-for-profit environmental education organization dedicated to preserving the world's oceans and improving the quality of life for present and future generations. Philippe and I heard him talk about it so often we could recite his mission statement from memory, but he finally got around to it in 1973. Here again, it was the getting around to it that mattered most of all; the ongoing operations and other pressing details of sustaining the organization were better left to others. Dad actually put us in charge, together with a new business partner named Frederick Hyman. (In the early 1980s, we launched a sister organization in France called Fondation Cousteau.) And then he left us alone, to "do our thing," and thereby make the planet a better place and open our unique window to Dad's silent world.

In many ways, the success of the ABC television show made the Cousteau Society possible, because Dad finally had the money and the reputation to endow his lifelong vision—namely, to help millions of people understand the fragility of life on what he liked to call our "water planet," and to have me develop programs to help educate children as future ambassadors of the sea. To this end, he granted the Cousteau Society exclusive rights to all of his work, thus ensuring that there would be a steady income stream to fund the effort going forward. Unfortunately, the arrangement also meant that Dad would once again be strapped for personal cash, even though his expeditions were fully and adequately funded.

On a personal level, I embraced this new aspect of Dad's professional life, because it opened up for me a professional life of my own. Happily, the Cousteau Society was almost immediately successful, signing up tens of thousands of members in just the first few years, at one point reaching as many as 160,000 members. Our success would be short-lived, as I will soon detail, although the organization continues to operate as I write this—in a different form. As we made our first tentative steps, however, I was also developing something of a reputation as a museum designer and was soon commissioned to design other exhibition spaces in California—and even in Texas. And I found, almost by chance, that I had a particular interest in education, so I devoted myself to our various initiatives with school groups and youth organizations. As a kind of offshoot enterprise, I started an organization called the Living Sea Corporation, dedicated to teaching students and educators about the sea. Early on, we developed a teaching partnership at Pepperdine University that soon proved so successful that the United States Army asked us to adapt our program for military personnel, which we ultimately did at a base in South Carolina.

And so I counted my early work with the society as a great and resounding success, most especially the opportunities it afforded me to work with children, teaching them about the sea and its inhabitants. As I embarked on this exciting new aspect of my career, I could not help but think of my father and his important legacy. His love of the sea, his spirit of adventure, his thirst for discovery. These were deep and profound gifts that he had passed on to me and my brother in a kind of micro way, even as he passed them on to future generations the world over. It was, and remains, a brilliant inheritance. We never discussed it in quite these terms within our family, but I believe we were all fully and constantly aware of our rich legacy. Deep down, it was a part of who we were and what we had become.

Even my son, Fabien, picked up on it, when he was just a small boy. Fabien told me of a Christmas visit to Sanary; he was about six or seven, he figured, because his sister, Celine, had just started walking. One of Dad's gifts to his grandson that year was a wooden dinosaur skeleton, and Fabien recalled many long, wondrous hours at my father's knee, listening to Dad's reimagined stories of how the dinosaurs lived and how they died. As ever, Dad approached his material as a romantic, not an academic. His goal was to entertain, illuminate, transport, not to lecture. It was an approach I remembered all too well from my own childhood, when Dad would collect me and my brother for one of our middle-of-the-night "lessons" about the constellations—for in truth, these were not lessons in any kind of traditional sense, but grand, enlightening adventures laid out for us in ways we could not help but understand.

Yes, my father had introduced the world to the grand, enlightening adventures of the sea; yes, he had helped to develop the "sport" of diving and had opened up the oceans in such a way that people could swim like fish, unencumbered and free; and yes, his cameras had recorded the spectacular sights and sounds of the deep and brought them into our living rooms. Such boundless wonder all from one man. And so I took it upon myself to share these precious gifts with as many children as I could, to open up the seas for them in the same way my father had opened them up for me—for all of us.

In addition to my work in education, I was also spending more and more time overseeing postproduction on Dad's shows and documentaries, because I was now based in Los Angeles where I was making good and important strides in learning some of the business and technical aspects of film and television production. I was also quite proud of the work I'd done on the *Queen Mary*, which was completed around the same time. The mini-theater we built to highlight Dad's films was an impressive, state-of-the-art design, and the museum space made important use of the ship's original specifications. It was all of a piece. Mostly, though, the project was successful for the way it helped me to make a home for myself in California and to set off on a career of my own. I was still carrying out my father's work, in a way, but I was bringing my own talents to the enterprise and pursuing a passion that was mine alone.

Philippe, too, had slowly managed to carve out a distinct place for himself—also in a once-removed sort of way, at first, in relation to our now-famous father. To an outside observer, it might have appeared that Philippe was merely a supporting player, a younger, more idealized version of Captain Cousteau than the captain himself, but Philippe had a singular set of talents. He was an accomplished diver, which of course was to be expected of a "son of Cousteau," but he was also a brilliant photographer and camera-man, lending his unique vision to Dad's work. He was a tireless and peerless organizer, bringing a much-needed sense of order to JYC's increasingly complicated expeditions. He created his own com-pany, Thalassa, through which he hoped to market his considerable talents as a filmmaker. Over time, he became a pilot as well, in this way accenting Dad's accomplishments as a naval officer and giving our family an entirely new perspective on the skies to accompany our handed-down perspective on the sea.

And he would do far more in the air than fly small planes. He also flew balloons and gyrocopters—sometimes, to mixed results. A gyrocopter, or autogyro, is a tiny, single-passenger airship that looks like a cross between a moped and a helicopter. Philippe just loved to hover around in the air and survey the landscape, but on a visit to Easter Island, he caught an unexpected draft of wind and ended up crashing into the side of a hill, breaking his knee. Some years later, I found myself in the South Pacific and met one of Philippe's old acquaintances, who for some reason had stored the bits and pieces of my brother's busted-up gyroscope as a kind of keepsake—I guess because when a "son of Cousteau" alights on your property amid shards of bent and broken metal, you are inclined to think of your find as a souvenir.

Dad was especially proud of Philippe's pilot license, because of his own youthful dream of flying. Just as the car wreck JYC had suffered in his youth had gently guided him to the sea, so too had it kept him from the skies, and he could now live vicariously through the accomplishments of his younger son. However, Philippe had something of a reckless streak. He could be impulsive, impetuous. These were traits that had no place underwater. There is no room for carelessness in diving. Each gesture and each decision must be carefully considered. It requires absolute precision and certainty.

The same could not be said of flying, and here Philippe's personality seemed better suited to the task at hand. He was intrepid, charming, brash, bold, and imaginative. He was by all accounts an outstanding pilot, but he was regularly in and out of danger. Underwater, you have no choice but to heed the risks and hazards that present themselves; in the air, however, a fearless pilot might be inclined to fly through a trouble spot. That was how it was with Philippe at the controls. Early on, he nearly drowned while trying to capture aerial footage of blue whales. Some years later, piloting the *Calypso*'s hydroplane, he "landed" on a cactus and could not sit comfortably for several weeks afterward. And once, during a jump, his rip cord broke, and he landed in such a way that his heartthrob, playboy face that so charmed television viewers was permanently scarred.

Very often, Philippe and I would remark on how our lives had changed, since growing up at each other's sides. In many ways, we were the same; in many ways more, we were not. He was fearless; I was somewhat circumspect. Certainly, we were bound for all time by our parents, by our experiences; we shared the same blood, and the sea ran through that blood in a deep and fundamental way. Now we were off doing our own thing—still in our father's powerful orbit, still circling around La Bergère, but moving to some of our own rhythms at long last, even if those rhythms were now subject to public scrutiny.

As boys, we had been each other's co-conspirators and constant playmates. Even when we were sent off to boarding school and could no longer interact in such an intimate, round-the-clock way, we enjoyed a special bond. As adults, set down along our separate paths, we managed to come together on board the *Calypso* at every opportunity. Philippe was married, too, with a young child of his own, and I often thought back to our charmed boyhoods, caught in the constant swirl of unpredictable excitement that seemed to follow JYC and Simone like sunshine. Whenever I closed my eyes and pictured the two of us together, we were splashing about in the waters of the Mediterranean, making mischief of some kind or other, or jetting off on Dad's whim to join the *Calypso* on a break from boarding school.

Now, as adults, we put our similar backgrounds to very different use. We lived miles away from each other, rarely on the same continent, but we managed to meet with some frequency. There was always a good reason to get together, even over such vast distances. We were so different in so many respects, and yet at bottom we were more alike than anything. There was no denying our common bond. Whenever possible, we would often meet at the Villa Baobab in France for breakfast. He was altogether a great, great guy, but he could also be unbearable. He was the only one of us to stand openly against my father if he believed my father was in the wrong. My mother and I would invariably choose a more subtle, more sidelong path and attempt to persuade my father to reconsider the matter, whatever it happened to be, but not Philippe. He'd turn to my mother and say, "Your husband is a fool!"

A son of Cousteau? In all the world, as far as I knew at the time, there was me and one other, and so it was a particular blow when I received word that my brother had had an accident. I could not believe the worst. In fact, I refused. Philippe's plane had gone down in Portugal. It was June 28, 1979. I was on the island of Catalina in California, teaching kids about some aspect of ocean life, which was ironic because Philippe had been flying a Catalina plane—a twin-engine amphibian model he had christened *Flying Calypso*. This was no tiny gyroscope, as before, hovering at just a few hundred feet; this was the real deal, flown at real altitude. Eight people were on board, including Philippe's copilot, when the plane broke in half as it landed on the Tagus River, near Lisbon. The passengers and copilot were quickly found by rescue workers, but one of the propellers had broken off and killed Philippe. A search mission was underway at the time to locate his body.

My mother was on board the *Calypso* with Albert Falco, our devoted Bébert, sailing down the East Coast of the United States toward Jacksonville, Florida. JYC was in New York, trying to get sponsorship for his next expedition. He'd gotten the same call as I had, but together, we determined not to tell La Bergère until JYC could be at her side. It was a devastating piece of news, and he did not want her to have to hear it alone. As we made our way toward each other, we clung to a small piece of hope that somehow, somewhere, Philippe might be okay. After all, his body had not been found; the others had survived, we soon learned, the copilot by some miracle, because the propeller had severed his seat belt and he was flung to safety. In the meantime, Dad instructed Bébert to keep the radio tuned only to music. "Madame," he said to my mother, when she looked to turn the dial to a news station, "if you don't mind, I would very much like to hear some more music."

Even so, Bébert could not keep La Bergère from the news indefinitely. Already, he had passed an entire afternoon in this way, and it would be some time before my father could arrive. Finally, with great hesitation, Bébert followed my mother to her cabin. "Madame," he said, "you must listen to me. There is a serious problem."

Later, Bébert would write that my mother seemed to know as she turned to him. Her face looked pale and haggard. She said, in a low voice that did not sound quite like her own, "Don't say anything. I know. It's Philippe."

And in this way she knew. Without anyone having to tell her. Without hearing the news on the radio.

I flew to Lisbon immediately, arriving around the same time as my parents. Philippe's wife, Jan, remained home for the time being in the United States with their daughter Alexandra. (Jan was also pregnant, as it turned out, with Philippe's son and namesake, who was born several months after my brother's death.) She would join us in Portugal in time for the funeral. It was a tearful reunion. We were received by representatives from the French consulate, who could not have been more welcoming or understanding. It was such a sad, difficult time. There was still no conclusive news about Philippe, which in itself was not good. However, this also meant there was still hope that he might be okay. For the first day, we could get no additional information. There was a violent storm, and visibility was terrible. We struggled to piece together what might have happened. All the time, as we waited for a word, my father was pacing, pacing, pacing. My mother sat silently, preferring to be left alone, lost in her thoughts, while I went back and forth between the two of them.

We knew, and we did not want to know, all at once.

Finally, on the third day, we received the call we had been dreading from the authorities, informing us that Philippe's body had been found. The propeller that had saved the copilot's life had cost Philippe his, careening into his back with devastating force as he was trapped in the cockpit.

The light that had gone out in my mother's imagination was now darkened for all time.

We were shocked and saddened and grieving, each of us in our own way. Jan, of course, was devastated; she had been building a life with my brother, and now that life had been shattered. My father fairly shut down. He closed himself in his room for the longest time, muttering to himself as he pressed the door to its foundation. I could not be certain, but it sounded to me like he was saying, "It's over. It's all over." Over and over again, like a mantra. My mother, in contrast, was tough. JYC could not bring himself to go and identify the body, so he asked me to go instead. I was heartbroken but trying to be practical. So many people were moving in and out of the morgue, with and without credentials, I could only think that one might pinch my brother's wedding band, so I slipped it off his finger. At just that moment, nothing seemed more important to me than being able to return that ring to my sister-in-law—although in truth the ring was just a token, a keepsake.

It would not bring my brother back.

I could barely bring myself to look. Philippe had been damaged so badly in the accident. But of course I looked. I had no choice but to look. And then I looked away. I said, "Yes, that is my brother."

Then I went back to the home where we were staying to tell my parents the edited version of what I had seen. "There can be no mistake," I said. "It is Philippe."

At this, my father broke down all over again—as though there had still been the tiniest sliver of hope; perhaps he had been thinking that the authorities had somehow discovered the wrong body. But, no. It was his son. He fell apart. He fell into muttering again: "It's over. . . ."

Philippe was honored with a burial at sea by the Portuguese Navy, and as his body was being submerged, his widow took the ring I had given her and threw it on his casket. It was a very moving, emotional moment—even now, all these years later, I can close my eyes and picture it.

My mother was tough, though, as I have written. So tough, she surprised me, just as she surely surprised herself. It was only later—weeks later—that she finally broke down herself and let her emotions show. But she kept those emotions in check at first, knowing at least on some level that she needed to be strong for my father. It's possible that her grief was so great she could not bring herself to cry, but I choose to believe that she remained stalwart for my father and I. Of the two of them, he was certainly the weaker in this regard.

PHILIPPE

There followed a surprising and emotional exchange between my father and me that began a new phase in our relationship— professional and otherwise. That very afternoon, my father turned to me despairingly and said, "Jean-Michel, if you don't come to help me, I'll abandon everything."

I did not hear it as any sort of threat or ultimatum. Rather, it was a plea, a declaration. A desperate cry for help. He could not go on without his son. The great Captain Cousteau was at a loss. Over the past decade, Philippe had become not only the eyes and ears of my father's operation, but its heart and soul as well. My father was almost 70 years old, and he did not have the energy to monitor his own interests and activities; as much as anyone else, Philippe had knitted together JYC's adventures in such a way that they appeared as a seamless enterprise, and my father was now all but pleading with me to assist him going forward.

Once again, I did not have to think about it. He needed a son at his side, so I would be that son. All my father had to do was ask, and I was prepared to drop everything.

I said, "Don't worry, JYC. I am right here for you."

I did not realize the extent of my pledge to my father at the time, but I was true to my word. I ended up having to sell everything and to shut down my various sideline pursuits, many of which had only recently gotten underway. We had bought a fine piece of property on Hilton Head Island, in South Carolina, as just one example, where we were planning to move to be near to our Army educational program, but we had to sell it at a loss. I had agreed to take on another museum design, but I had to leave the work to someone else. The one thing I would not give up was my educational program, which I insisted be adopted by the Cousteau Society, but there was never a question in my mind: I would do whatever my father asked of me.

Philippe's tragic death left a hole in all of our lives, and yet, as we settled into our grieving, it appeared that my father was facing the biggest hole. This is not to diminish the pain and suffering of Philippe's wife, Jan, the depths of which I cannot begin to imagine. But my focus here was on my own family, the family Philippe had known as a boy, the family he had now left to me. For whatever reason, my father could not see his way to the other side of that hole—and, right or wrong, I vowed to help take him there.

FUTURE GENERATIONS

There is an old proverb that was a particular favorite of my father's: *L'avenir peut être à nous.* It is a play on words in the original French. In translation, it is more straightforward and simply states that "the future can be ours"—a hopeful perspective for such a confirmed skeptic as Captain Cousteau.

Either way, it puts me in mind of how my father approached his life and work and the shift in perspective that found him over time. Dad was not typically given to thoughts of mortality, although in the years before Philippe's death, he had begun to speak of the death of our natural resources and the declining health of our planet. He was particularly concerned about the future of our oceans, which had lately become polluted, "fished out," and hardly resembled the gracious plenty he had stumbled upon as a young man. Today, it has become the norm for ecologists and conservationists to sound the alarm on environmental issues, but as recently as the 1970s, there was no such hue and cry—certainly not on any kind of grand scale.

Agree with him or not, my father was one of the earliest and loudest proponents of the "green" initiatives we now take very much for granted, and yet because he was not a scientist and merely a popular adventurer, his concerns were generally dismissed within the environmental community. His work might have given him a voice, but on this he could not get a substantive hearing. This did not keep him from stepping up his efforts in this regard in the months following Philippe's death—and for me, the skeptic son of the skeptic captain, the words of Dad's proverb began to ring especially hollow.

The future can be ours?

Perhaps, I thought at the time. Perhaps not. We shall see.

My skeptical father was not prone to thoughts of legacy or posterity, either. It was not in his nature. He was certainly pleased that he had made a positive contribution on the world stage and that he was continuing to do so, but he did not worry much about how he would be remembered or even if he would be remembered at all. That is, he never had done so up until Philippe's death. Now, my father seemed to consider his own mortality all the time—or, at least, with some frequency. I started to hear him speak of life and death as two sides of the same coin; it became a common theme for his diatribes and musings and a central concern. He also spoke more and more of his own father, Daddy, who lived to be 92, keeping active and connected and vibrant until the very end of his life. That was how JYC wanted to leave this Earth; he wanted to remain relevant, in what ways he could, for as long as he could. He wanted to matter, in the way he had mattered for the past half century. Just then, nothing was more important, and here his cherished proverb seemed to carry an alternate meaning. Here the "future" was about the legacy he would hand down to his children and his children's children; it was about the world he would leave behind.

It was in this context, then, that my father came to me one afternoon not long after my brother's funeral to discuss yet another matter of some urgency—only this urgency was decidedly (and, uncomfortably) more personal in nature, as I would soon learn. As further context, I believe it helpful to note that JYC had long been rumored to have had his share of extramarital affairs. In this, I suppose, he was not unlike a great many European men of the period—a great many men of the wide, wide world, I'll allow— although it was a difficult prospect for a son to consider. To my mind, my parents had enjoyed a singular love affair. Theirs was in many ways a model relationship. This was not an idealistic view, I don't believe. JYC and Simone truly seemed to enjoy each other's company and to appreciate each other's gifts and interests and points of view. They were and remained attracted to each other, even after all these years. Above all, they complemented each other, in the manner of lifelong companions capable of completing each other's sentences, sharing each other's dreams, and working toward a common goal.

Simone knew my father's shortcomings, and yet she accepted him as he was. And JYC knew my mother fully as well. He was undoubtedly the only man she had ever loved, while I must allow that she was most certainly not the only woman to have enjoyed my father's affections, even as Philippe and I grew up believing she would be the last. Let us be practical: My mother lived most of the year on board the *Calypso* and was thus almost beyond suspicion; her comings and goings were well known to all. In contrast, Dad's comings and goings were often known only to him. He came and went as he pleased. He was a man of the world, whose reflexive and seductive charms were a key aspect of his character, among men and women alike, and was thus almost inviting suspicion—especially now that the apparition of celebrity had crept into his life and seemed to add to his already considerable appeal. Always, he had measured himself against the affection, admiration, adulation he received from others, and following his success in television, he was taking in more than his share.

And yet it was none of my business. What passed between two people in a marriage was theirs and theirs alone. What passed outside a marriage was also theirs alone; it did not even belong to their adult children, who in many ways had a vested interest in knowing the truth. This was how I approached the whispers and suspicions that invariably found me regarding my father's alleged infidelities. I chose to ignore them, even when they were difficult to ignore, like the time when my father came to collect me at the airport in Los Angeles, driving a sleek yellow MG convertible that was definitely not his own. I knew his taste in cars. I knew he would never choose such a dainty vehicle. And so I put it to him, as I eased onto the passenger seat.

I said, "What on earth is this, JYC?"

He dodged the question with a wave of his hand, but he must have known he'd been found out. And he surely was: The car belonged to a famous Hollywood actress, with whom he'd been carrying on for a number of years. Anyway, that was the rumor. I'd seen her driving around town in this very same car, so I had to think that Dad must have wanted me to know about the relationship in some way. But I left it at that, and the question hung in the air between us like fumes.

Therefore, it was with some measure of surprise and a great deal of confusion that I met with him for lunch to discuss his "urgent" personal matter. We had not seen each other for some time, having been off in our separate swirls, on separate continents. Also, and significantly, we had not really talked openly about Philippe since his tragic plane crash, other than to redirect my career plans so that I could now attend to the details and formalities of my father's work and that of his foundation, but that was not unusual for us. Our emotions could often be found in what was left unsaid, because the men in our family did not always talk openly about such matters. We talked around them instead—only here I began to feel there was a bit more on the menu for this particular lunch. As was our custom, we met at a restaurant on the Boulevard de Courcelles, close to the Villa Wagram in Paris. We were quite comfortable in these surroundings. We talked about our now-intertwined professional interests, about upcoming expeditions, about the Cousteau Society, which had quickly fallen into disarray, about other projects we might consider in the future.

Ah, the *future*. There was that word again, never far from the surface for my father now that Philippe's death had forced JYC to face his own mortality. We ordered a nice bottle of Bordeaux to lubricate the silence that fell across our table now that we had gotten past these few pleasantries. It was obvious to me that Dad had something on his mind, something that was troubling him. He could not think how to begin. Perhaps it had to do with his finances, I thought at first. Perhaps he had a problem with some of the decisions I'd been making regarding our foundation. Perhaps it was something else entirely.

Finally, he started in on it. "Jean-Michel," he began, "I have to tell you. . ."

Then his voice trailed off. He spoke in an uncharacteristically low voice, almost conspiratorially, and then he stopped short, as if he was afraid to complete his thought. Whatever he had to say, he was no closer to it than when he'd started out.

I took another sip of wine and waited for the silence between us to pass.

He started in again: "Jean-Michel, I have to tell you. . . "

Once more, his voice trailed off. He could not seem to find the words, and I could not think how to help him, so once more I drank. This time, he did as well. At another time, we might have lifted our glasses in toast, but what was there to celebrate on this occasion? Very little, since losing my brother. We could not even celebrate the birth of Philippe's son and namesake, born just a few months after my brother's death. Indeed, the birth of her grandson Philippe had been a time of bittersweet anguish for La Bergère, who finally broke down. For months, she had kept her emotions in check, pretending a strength in support of my father's weakness. I suppose the sight of a tiny, brand-new version of Philippe brought to mind what she had lost, what we all had lost.

It was a solemn time in our family, and so we merely drank.

"Jean-Michel," my father tried once more. "I have to tell you I am having an affair."

On hearing this, my first response was to try not to laugh. I did not wish to be coarse or unfeeling, but this was no revelation. Rather, it was like remarking that he had recently drawn breath. After such a show, to come out with a confession that was glaringly apparent to all concerned and unconcerned alike was, well, laughable. Still, I set his confession aside and considered it. Maybe this "affair" would do him good, I thought. May he not be so full of himself as to think I must be made to consider his every move. Indeed, I knew of the woman who had lately captured his affection—and she was 35 years his junior. I could not see where JYC was going with this conversation and could not imagine that it was any place I wished to accompany him, so I put up my hand, as if to hold off whatever might come next. But I was too late. My father had gathered a kind of momentum, so he leaned in further still and continued: "She is expecting a baby."

Suddenly, I could not laugh. What had briefly appeared before me as a developing comedy was now a fully formed absurdity—and, an affront. To hear a man, several times a grandfather, confess to his by now middle-age son that he was having an affair with a woman 35 years younger than he, a woman who would shortly bear him a child—it was as if the skies had opened and swallowed my world whole. (My goodness, his "girlfriend" was younger than I was.) I could close my eyes and picture a torrent of waves, crashing down upon our precious *Calypso,* washing over years of love and teamwork and family history. *My* family history. And sinking us all! This was surely the most terrible confidence my father could have shared with me, and it only got worse, because once he shared his secret, a great weight seemed to lift from him and pass on to me. He appeared liberated—free—while I felt burdened, conflicted, bound by a filial sense of duty to my father and my mother.

I realize now with some compassion that it could not have been an easy thing for the great Captain Cousteau to confess such as this to his firstborn son, but he followed his confession with a troubling observation that continues to rankle 30 years later. He said, "The thought of this child is a great consolation after Philippe's death."

I cannot be certain that he meant it in quite this way, but what I heard was that in JYC's mind, the child he was expecting would replace the child he had lost. It was a simple equation and a shocking admission. I was outraged, confused. I thought immediately of my poor mother and wondered what she would make of Dad's senseless remark—and hoped he'd have sense enough to keep such thoughts to himself. It was difficult enough to imagine how Simone

217

would react to my father's transparent infidelity and to the baby itself, but to hear him speak of this unborn child as a replacement for Philippe would have been unbearable. How could one child ever replace another? To even give voice to the idea was an abomination, and yet my father showed no remorse over his remark. In fact, he continued to make his point, perhaps to bring me around to his way of thinking. He said, "It is true, Jean-Michel. The news of this pregnancy has lifted some of the pain I've been feeling. It's filled some of the emptiness. I am a new man."

I reminded my father that he was not a new man at all, but rather a fairly old one, steeped into relationships and routine. Also, he did not need me to point out that he had been a life-long proponent of birth control, to curb the overpopulation that plagued our planet. "We will be suffocated by cradles!" he used to say. But I pointed it out to him anyway, and now here he was preparing to add another cradle to the mix—quite happily, it appeared. Had he forgotten everything he'd been preaching for the past half century?

Despite his confession, it was clear to me that he remained very much devoted to Simone. She was his mentor, his spirit-guide—his maypole, as well as ours. But there was room in his heart for another, apparently.

I could not think what I would say to my mother, although I feared I would have to say something. But when? And, how much? I feared, too, that I would have to meet my father's "mistress," who was now a part of his life whether I approved of the relationship or not. After all, I would be an older brother, of a kind, to the child she carried. I could not stiff-arm that relationship any more than I could push my own father from my life. There was no profit in making my animosity known—if indeed that's what it was. However, there was no denying that our meeting would be a kind of charade. Whatever I chose to say to my mother, or not to say to her, would be another kind of charade.

What would I say to her, if she thought to ask? She was a perceptive woman. She knew, all along. On some level, she always knew. And there would be one day when she would come straight out and ask me, "Jean-Mine, do you know your father's mistress?" Or, "Do you know if there are any children?"

But then I realized that she would never put me in that position. She had not endeavored to do so in all of my adult years, and she would not do so now. It was her way of protecting me. And, in her fashion, it was her way of protecting my father—or, at least she was protecting the man she hoped he still could be. And so we never spoke of it. As far as I ever knew, she never spoke of it with my father, either. Surely, she must have known. But she set it aside and found a place for it just below the surface of her emotions, where it could not help but color her emotions for the rest of her life. Looking back, I could appreciate my mother's occasional mood changes for what they were. She appeared to turn inward, following this latest turn. She had always been a woman of great good cheer—only lately there had been times of brooding, even before Philippe's death. She would often retire suddenly to her cabin, without explanation. She would become quiet and introspective for reasons that were never readily apparent. I'd never thought anything of it, but now I took this behavior as a sign.

She knew, I convinced myself. Of course, she knew. And my father had to have recognized this, I came to believe. He, too, was perceptive. He was not only a keen observer of the sea and its inhabitants but also a student of human nature as well. Simone never questioned him about his long absences away from the ship, away from her, even as these absences grew longer and longer and took place with more and more frequency. But the simple, graceful act of not questioning his behavior should have given my father all the answer he needed. He should have known, and I was furious with him for creating a scenario where my mother—and, now, me!—would become complicit in his transgressions by virtue of our unwillingness to confront them.

To confront him!

And so we continued on in my father's charade, saying nothing, playing at normalcy. For years, we kept it up. La Bergère remained stoic and dignified, refusing to acknowledge what was soon plainly obvious—refusing to give up an ideal to which she had devoted a lifetime. When the child arrived, I dutifully visited my father and the child's mother and tried not to make my displeasure known.

In his own way, Dad participated in the same charade. For the rest of his days, he would live two lives, in two worlds—with two hearts, I have long suspected. And here is a curious thing. As he transitioned from one life to another, from one family to the next, from middle age into old age, I began to notice a change to his worldview. This change was almost imperceptible at first, and yet it gave color and meaning to the final "act" of his life. If it is true that a man lives his life in three acts, then it's possible to suggest that my father's denouement brought a kind of idealism to his work that I had never before seen. There was a passion, an advocacy, a purpose, where there had once merely been abandon and wonder and the pure pleasure of discovery. Suddenly, the future he had so long derided—the future of the planet, the future of his children and grandchildren, the future that might determine how he'd be remembered—was now all he could think about, and the French proverb he continued to quote took on yet another meaning. By the end of the 1980s—nearly a full decade into the "half life" he shared with his new family, which now included a second child, Pierre-Yves, to accompany the first one—he could finally be called a fierce battler for the future of our planet. What had taken him so long? Who can say?

He began to speak out about the diminution of our fresh-water resources—using his powerful platform as the self-anointed ambassador of the seas to put forth the slogan "Every drop counts," and hope it might make a difference. Perhaps it did, in the end. In any case, he was one of the very first to warn us that our drinking water might become scarce, even in countries where it seemed to flow freely.

"Despite all its science and technology," he wrote, "the human species has no means to satisfy its thirst for fresh water indefinitely."

It was, he said, our greatest concern. He turned his attention to the continent of Antarctica, which he had often described as a kind of outpost of life on Earth, a place where life clings to life itself. The air is colder there than anyplace else, he noted; the water more frigid, the elements more extreme. He understood as well as anyone how Antarctica stood as the continent at the end of the world—a continent that played a vital role in the balance of our planet's climate, despite its "terrible harshness."

He paid particular attention to the so-called Wellington Convention agreement, which had recently been signed by representatives of 33 nations, opening the entire Antarctic continent to exploitation of its mineral resources. Dad considered the agreement a travesty and an abomination—"an onslaught on the planet," he called it in one of his published tirades. In rebuttal, he enlisted a great many of his conservationist allies, including polar explorer Paul-Emile Victor, to join him in a petition denouncing the agreement and calling for Antarctica to be transformed into a kind of world preserve, not only to protect its own environment and wildlife but also to protect the world's climate and the level of its oceans.

Now in his late 70s, Dad might have been getting older, but he was unrelenting. He made Antarctica a true cause célèbre. In his native France alone, he collected more than a million signatures, decrying the Wellington initiative. He made appeals to French President François Mitterand, and together we went to see Australian Prime Minister Bob Hawke. He organized a Franco-Australian moratorium in favor of his "world preserve" concept, taking it directly to New Zealand Prime Minister David Lange. JYC even sent me to Argentina, where I enlisted the support of President Carlos Menem. Such were the advantages of a life as an environmental ambassador and international television star; people were inclined to give you a hearing, especially when your voice seemed to echo public sentiment.

The future is ours?

JYC didn't think so. If he believed this at one point, it had seemed to slip away. He was an old man. He knew he was an old man. He could sound the call for preservation at full voice. He could collect millions of signatures and gain an audience with people of influence and consequence. But he no longer felt the world was his for the taking. It was, however, his for the giving. He came to this realization relatively late in life, but here on this particular front—the conservancy of the continent of Antarctica—he put this realization into action. He recognized that his appeals and petitions were mostly an opportunity for him to vent and to make his position known. It would take a smaller, much more intimate initiative to truly make an impact on this issue, and I offer it here as an illustration of how we can often paint a big picture with small strokes. The expression works doubly well in English, because it calls to mind the simple beginnings of my father's work with les mousquemers and his growing reputation in those early years. His first "splash," if you will, came on the back of the small strokes he was making in the waters off Toulon—quite literally with his primitive mask and fins—and here he finally found a way to call attention to the Antarctic in simple, elegant terms.

And so, frustrated at the lack of movement from world leaders who had received him openly and heard his concerns, Dad decided to embark on an awareness mission with six children, selected from six corners of the globe in 1990. It was an ingenious concept, I thought at the time. Even today, I consider it one of his greatest accomplishments—an enduring tribute to the man and his work, embodied in the hearts and minds of the world's children. I believe it also offered a glimpse into his own emotions, as he gathered all these different children, one from each continent, in a way that could be seen not only to mirror the human condition but also his own circumstance as well. After all, he had a working lifetime to conceive of a project such as this—why now? Why wait until he had fathered a "second family" to think along these lines? Why surround himself with so many children, about the same age as his own young children?

Yes, the subject was ostensibly the looming devastation of a continent with a landmass 25 times greater than his native France. Yes, it was an opportunity to reveal Antarctica's beauty and its fragility as well; it is, after all, an extraordinary landscape, harboring animals that are extremely vulnerable to human-induced changes. And yes, atop Antarctica's legendary ice shelf, he could show footage of penguins milling about in colonies, while "underneath," his cameras could capture seals playing just below the surface. There could be no denying that the continent offered a marvelous canvas. There would be all manner of fish, squid, crustaceans, and sea stars living in these freezing waters. Dad's cameras would surely soak up the spectacular scenery. In announcing this expedition, he called the Antarctic region "an inestimable treasure that we must preserve intact for future generations" and asked that the continent be declared "a natural reserve, a land of peace and science."

If there was a successful formula for a Cousteau expedition, then this one surely fit the bill. There were a great many arguments for visiting the region at just that moment, but few for doing so with all these children in tow. The idea was to shine a light on the Antarctic continent through the eyes of these six youthful ambassadors, who ranged in age from eight to ten years old. There were three boys—from Tanzania, Chile, and the United States—and, three girls—from Australia, Japan, and France. They were accompanied by a young female chaperone from the International School of Paris and sailed with my father and his team on a two-week voyage aboard the *Erebus*, a ship he had leased for just this purpose. The children served as the crew—in age-appropriate ways. They were meant to represent the eyes and ears of the world, as they took in the potential devastation of the continent, from a place of innocence. But in some ways, I believe they were meant to stand in for all of my father's children—biological and otherwise. In a deeply personal way, they represented the children he would leave behind, the legacy of adventure and wonder that would be his precious gift to the world.

It was an altogether inspired expedition—and I'm happy to report here that it succeeded mightily. The film that came out of it, *Lilliput in Antarctica*, was one of my very favorites for the way it coalesced my father's vision of the present with his vision for the future.

I had the good fortune some years ago to reminisce about this mission with one of its young participants. Elise Otzenberger was now a young woman. She had joined Dad's "future generations" expedition with a gentle push from my mother, who knew Elise as the daughter of Claude Otzenberger, the noted television producer and director of a number of Cousteau films. Elise grew up around the water, around the *Calypso*. By age eight, she was an accomplished diver, in much the same way Philippe and I had taken up the sport as boys mimicking our father. Somehow, young Elise developed a fine friendship with La Bergère—the lady of the *Calypso* and the little girl of the sea.

Simone listened in as Dad and his partners were discussing which children to take with them on the voyage. There were hundreds of children to consider, each more talented or qualified or curious than the next. It did not seem that they could make a bad choice in their "casting"—as indeed that's what it was. Finally, my mother weighed in.

"What about the little girl?" she asked.

JYC looked up from his notes with a measure of his own curiosity. *Which little girl?* his eyes seemed to ask.

La Bergère knew my father's questioning eyes, so she answered before he could speak. "Elise," she said. "Claude's daughter. She's marvelous."

She certainly was. Dad could not imagine why he hadn't thought of her himself, although of course he was trying to assemble a representative coalition of children to symbolize our global population. It simply did not occur to him to reach out to a child he already knew, but now that Simone had put Elise's name on the table, he sparked to it immediately.

"Of course," he said, smiling. "The little girl."

The little girl remembers the voyage as a highlight of her young life. She already knew my father, of course. At the start of the journey, she met with four other youngsters: Kelly, from Australia; Oko, from Tanzania; Cory, from the United States; and Fumiko, from Japan. Geronimo, the Chilean boy, would join the group a short time later, on a stopover in Santiago. None of the children spoke the same language, and yet they soon became friends, united by their shared adventure and this wisp of a man with a red woolen cap who seemed to overflow with warmth and good cheer—a man old enough to be their grandfather and yet young enough in spirit to be their friend.

"JYC's presence was very much felt by us all," Elise recalled when I caught up with her to talk about her experience. "He was always in such a good mood, never showing the slightest irritation, even when we squabbled or argued or did something stupid like little kids always do. He was interested in what we were thinking and in what we were feeling. In this way, he was very different from the other adults we knew. He never treated us like children. And when he asked us what we thought, it was not to flatter us or indulge us. It was because he really seemed to value our opinion."

All these years later, Elise Otzenberger remembers the glorious arrival as the *Erebus* reached Antarctica, escorted by a pod of whales that seemed to welcome them to these waters.

My father's notes from his log make special mention of this moment:

January 6, 1990. At the gate of Antarctica, the *Erebus* enters Admiralty Bay at reduced speed. Blocks of ice are scattered in the bay. On the bridge, our six little ambassadors let out shouts of joy, as two pairs of humpback whales swim about the ship, disporting themselves in a show of welcome. It is a marvelous sight. Six little faces of varying colors. Six sets of wide-open eyes. Six children, off to reclaim the continent of Antarctica in the name of all the children of the world, those of today and those of tomorrow.

The following morning, the children discovered the skeleton of a blue whale, which saddened and fascinated them. The bones were remarkably well preserved, owing to the cold. Dad explained that all materials in the area would deteriorate slowly in these extreme temperatures, including a small plastic bag or any other pollutant and contaminant, which was why it was so important to reduce or eliminate any industrial waste from the region.

Then, at the mouth of Admiralty Bay, they came upon a colony of penguins—the emblem of Antarctica. (And the mother of all photo opportunities, for a documentary filmmaker.) Soon it was decided that the children would get out and build an igloo—with some help from Dad and his team (another powerful visual). However, as Elise remembers it, the children were the principal architects and haulers, eventually carrying blocks of ice as big as themselves. Upon completion, the igloo measured 9.8 feet in diameter and 6.8 feet in height. Kelly and Oko volunteered to sleep in it the first night, thereby taking symbolic possession of the continent in the name of future generations.

The expedition had as deep and profound an impact as any my father had undertaken in his career, and yet I could not look on without thinking of the special resonance the trip seemed to hold on a personal level. For me, for my mother, for my father—and, now, for his "second" family as well. I do not believe he ever drew the same conclusions about this undertaking as I had drawn or found the same points of connection as I had found, but they were surely there just below the surface—which, after all, was precisely where he kept his emotions.

Lilliput in Antarctica was broadcast to great reviews and ratings on Antenne 2, the French television network, as well as in the United States on Ted Turner's TBS network and other networks throughout the world. To capitalize on the positive attention, Dad also made one thousand copies of the film and sent them to every American senator and congressman, as well as to the justices of the United States Supreme Court. He was determined to spread the word on behalf of the Antarctic region, to let the innocent eyes and ears of his "future generations" delegation give full expression to his cause. Subsequently, he was able to collect an additional three million signatures in a petition and persuade one-third of the original Wellington Convention nations to vote in favor of a total ban on any type of mineral exploitation on the continent. It was a great victory. Dad even met with President George H. W. Bush in June 1991, to press the matter further. I happily joined him in Washington for this meeting and marveled at the way the captain set about getting what he wanted. The following day, the President announced that the United States would join in signing the Madrid Protocol agreement, declaring Antarctica a "natural reserve and scientific territory."

Against these developments, then, I considered my father's cherished proverb yet again: *L'avenir peut être à nous.*

The future is indeed ours. JYC came to believe this wholeheartedly. But he also believed it belonged to our children—to the world's children, of course, but also to his children and grandchildren in particular.

"What kind of planet are we going to leave to them?" he often wondered.

It was a question for the ages. And here was my father, in the twilight of his career, standing in answer.

THE END
OF SOMETHING

June 11th, 1990—a day of celebration, at a time in our
lives when there was not much to celebrate. Certainly,
in the decade since Philippe's death, there had been no real
joy in our family. We had been moving along on our separate paths, that is all. Whatever had knitted us together in
the past had by now fallen away. In that time, my father
welcomed two small children to his ranks, but there was
no place for them in our traditional family dynamic, just
as that dynamic itself was no more—or, at least, was not
what it had been. As far as I ever knew, my parents had not
discussed my father's ongoing relationship with the mother
of these children or even the fact of their birth, and my
mother never let on to me that she knew of any such thing.
And yet, I had to believe she knew. Oh yes, on some level

she knew. And my father, ever the student of human behavior, must have known in turn that she had figured him out as well.

And so the pretense continued, and we Cousteaus continued on in what was left of the spotlight that had once shone so brightly on my father's work and the glow that had once attached to our seafaring family. Even now, as JYC dialed down his expedition work in consideration of his age and the age of the *Calypso*, he was very much at the forefront of attention. The Cousteau Society, the foundation he'd established with me and my brother to augment his research, continued to attract new members all over the world. The arrival of cable television meant more and more outlets for Dad's films to be shown. A new generation of viewers was being exposed to Jacques-Yves Cousteau, which meant that in some respects JYC's work was more popular than ever, even as the man himself appeared to retreat from the public stage.

Don't misunderstand, my father remained active, but this is a relative term. He was active for his age and circumstance. He continued to sail and explore and lecture. He continued to write books chronicling his adventures and calling attention to the world's oceans. He continued to receive deserved accolades from world leaders and prominent environmentalists for his pioneering work. But as he approached his 80th birthday, he could not be nearly so active as he had been throughout his life. He would not give in to the ravages of time, but he certainly accommodated them.

My mother retreated, too. However, it appeared to me that she did so beneath the fog of alcohol and cigarettes. She had always been a heavy smoker, a habit that often ran to more than a pack a day. She also enjoyed her glasses of wine—she was French, after all—but in recent years, those glasses had become a bottle, and sometimes more, and she would often retire to her cabin in the middle of the afternoon, unable to hold up her end of a conversation. I would nearly become angry with her for drinking with such apparent success, but my anger would have been misplaced. You see, earlier that year, La Bergère had been diagnosed with cancer—an outgrowth of her heavy smoking, no doubt. Her prognosis was not good, and yet she gave strict instructions to her doctor, Denis Martin Laval (who used to be the ship's doctor on board the *Calypso*) not to tell anyone of her illness—not even her family. And so she kept her suffering to herself. Dr. Laval reported to me after her death that she didn't want her illness to get in the way of our work.

By the time her cancer was discovered, it was in its advanced stages, so Simone proceeded as if she was beyond treatment. Perhaps that was so. In any case, she continued to suffer in silence—an infuriating show of courage, I came to believe later. Moreover, she did not want it known among the *Calypso* crew, with whom she spent most of her time. She drank, I now suspect, to wash away the pain and the sadness and to fill the spaces where her young family had been. Where there was once élan and purpose, there was now only sadness and despair—and where she had once looked ahead to the wonders of the whole wide world, she could now picture no such thing. She was also in a great deal of pain, and because she was often alone in her pain, there was nothing to do but drink and endure.

My father, by contrast, was often at his apartment in Paris, preferring to sleep on dry land when the ship was at harbor. At least, this was what he always said when pressed on the matter. And so my parents were rather like two ships passing in the night—an apt metaphor for a marriage that had been so long at sea. In the decade and more since Philippe's death, they had been together only infrequently, which only facilitated the travesty of my father's "separate" family. His life and work were such that he was free to come and go, often as he pleased, and here it appeared that he chose this other woman's company over my mother's. Had he known of Simone's illness, he might have behaved differently, but as it was he kept himself busy and unavailable—to each woman, on some level, but to my mother most of all.

And yet there were glimpses of what we had all meant to each other as a family, what we meant to each other still. We came together on this day—JYC's 80th birthday, as it happened, but notable because my mother was being honored at the Oceano-graphic Museum of Monaco. We had looked forward to this day for months, ever since it was announced in the newspapers— although Simone never really read the papers. She preferred *Paris Match* or *Elle* for keeping up with French society, but this bulletin was called to her attention. La Bergère was to receive the Order of Maritime Merit—a notable honor, to be sure, one only rarely conferred on a woman. JYC himself brought the news to Simone, and he did so as if he'd had something to do with it. He came to her smiling, laughing, rejuvenated—in a manner she had not seen in him for some time. She told me later that he even addressed her as Loubi, a shared term of endearment she had not heard from his lips in years. In her head, she imagined they were young lovers once more. The gulf between them seemed to fall away. My father might have been thinking of her in the old tender way, but he was also bursting with excitement at the news he had come to deliver. He could not get the words out fast enough. At first, my mother thought he'd gone mad, the way he was talking so animatedly, so enthusiastically. "My goodness," she said, "he seemed terrifically excited." But about what? She could not follow his words. An offi-cial decoration? For her? He wasn't making any sense.

"No, no, no," he insisted, waving a copy of the official procla-mation. "It's true, it says so right here. The ministry wants to honor you for a lifetime of service rendered to the sea."

She could only laugh, my dear, long-suffering mother. She was not expecting such as this. She was used to deflecting the praise and admiration that invariably came her way, so she attempted a joke: "It is the sea that has rendered me a great service," she said. "After all, my dear JYC, without the sea, I would not have stood by you all this time."

My father was certainly amused. Even after all these years, after all that had passed between them—and, lately, all that had not passed between them—my mother could still make him smile. JYC was enormously proud to be the bearer of such grand, deserving news, even though he had not been behind it in the least. It would have been a natural assumption for my mother to conclude that he had played a role, but the fact was that the honor had come with a gentle push from my parents' great and loyal friend Alain Traonouïl, who had been captain of the *Calypso* for several years. He and my mother had been especially close, as seemed to be the case with most of the men who lived and worked on the ship. JYC, Traonouïl, Albert Falco, André Laban, Titi Leandri—to a man, young and old, novice and veteran, they were all devoted to La Bergère, in the manner of Peter Pan and his Lost Boys, sitting in thrall of their beloved Wendy. Here, the faithful Traonouïl reached out to the ministry on my mother's behalf—and then neatly stepped aside to allow JYC the honor of announcing the news. As was his custom, my father did not claim credit for making these arrangements, but he did not redirect it when it came his way.

However it came about, Dad was plainly overjoyed to be in a position to present my mother with such an honor. It was as if he were giving her a profound gift—one that had fairly fallen into his lap, but one he was happy to bestow just the same. He was deeply grateful for my mother's support and good counsel. And, difficult as it might be for some to believe in light of his protracted infidelity,

233

he remained devoted to her. He even needed her. And, she needed him. The one was bound to the other, each in his or her own way.

And so we went as a family to the Museum of Monaco, where Captain Jean Alinat presented the medal to my mother. La Bergère showed up in splendor, dressed in a silk Japanese gown that called to mind the traditions of her childhood. She looked radiant, magnificent. In a hundred years, I would have never guessed that she was ill. But she was surely suffering, and this fine occasion would mark one of the last times I would see her sparkle in quite this way. In truth, she had always resisted the glare of public attention. She much preferred jeans to silk. My father once told a reporter that she was happiest out of camera range, in the crow's nest of the *Calypso,* scanning the sea for whales. "She lives to spend hour after hour in the wind and sun," he said, "watching, thinking, trying to unravel the mystery of the sea."

She died just a few months later in her apartment in Monaco. The *Calypso* was in port in the Philippines, between expeditions. Had the ship been in her home port, La Bergère would have surely been on board. As it was, she waited for my father to return home from Paris for the weekend, and she quite literally died in his arms. I was in Cambodia when I got the news. I was devastated, but as I made arrangements to return to Monaco on the next flight, I began to recognize my mother's reclusive behavior of the past months for what it was, and my heart ached in entirely new ways. It ached for me and my father and the loss we would now feel; it ached for the ways my father had treated Simone over the past decade; and, it ached especially for my mother, because I could see that she had suffered alone. It may have been her choice, but it was a great sadness just the same.

Nevertheless, my mother's wishes were clear, and they ran to her final arrangements as well. Her ashes were lowered in the "trench" in the sea of Monaco, following a full military funeral service. It was a fitting tribute for a woman of the sea—and, in many ways, it was an everlasting gift to me, because it meant that whenever I was to visit these waters in the future, my mother's spirit would be near.

My mother's death almost meant the death of the *Calypso,* because I do not think we could have retired the ship if my mother had been still alive. No, she would never have allowed it. As it was, I returned to Cambodia shortly after my mother's funeral and rejoined our expedition. The *Calypso* was not with us on this leg, because the team explored the upper Mekong River with too many waterfalls for our old minesweeper to navigate, so we joined up with her later for the return trip downriver. At the mouth of the Mekong, those of us on board came to the conclusion that this would be *Calypso*'s last great voyage. It was not so much a conclusion as an inevitable progression. At the end of our journey, I went to Paris to discuss the matter with my father. The ship was no longer safe. Part of the stern was starting to break off. It was costing us approximately $10,000 a day just to keep her afloat.

I feared it would be a delicate, difficult conversation because so much of my father's life and work was invested in his beloved *Calypso.* Much of his relationship with my mother was rooted there as well, but he took a practical view. "That's it," he said. "She is through."

Here again, I believed this was another fitting tribute to La Bergère, because it meant the *Calypso* would never truly sail without her. Yes, the ship completed the Cambodian expedition that had been undertaken before my mother's death, but it was merely her last lap. She was only finishing what she had started, but it was as if the wind in her "sails" was gone. Of course, we couldn't just leave the *Calypso* and be done with her. She stayed on in Vietnam for about six months, with a small team that tended to her necessary maintenance, after which we decided to send her on to Singapore. There, as she was awaiting repair, calamity struck. By some tragically bizarre coincidence, a barge being towed by a tugboat careened into the *Calypso* at the dock in Singapore. The *Calypso* was made of wood and the barge of metal, and it tore right through her hull. She sank in about three minutes, right there in the harbor. She was later salvaged and put on a barge for Marseille, where she could sit proudly in her home waters as a kind of monument to my mother, Simone Melchior—the ship keeper.

The ship was a tribute to my father as well, but not in the same way, for the *Calypso* had not been his one and true home, just as my mother was no longer his one and true love. He was broken up over my mother's death, just as all were who knew her and loved her, but for JYC it marked a kind of liberation. Suddenly, he was free to shake the pretense from the rest of his life and pursue his relationship with the mother of his young children in a more public way. Soon—perhaps, too soon—they made plans to be married. My father sought me out one day to discuss his decision and to ask if I would attend his wedding. It was just a few months after my mother's death, and yet I did not begrudge him this new relationship, or a second chance at a second family. I did not even mind that he was planning to remarry—at least, theoretically. I might have resented his neglectful, unmindful treatment of my mother over the last decades of her life. I might have wished away his many transgressions over the years. But all of that was behind us. What mattered were the kindnesses we would offer each other, the space we would make for each other in our lives going forward.

JYC might have believed he was entitled to many loves, many children, but I would only have one father, and so I agreed to attend his wedding. I did so with a heavy heart and an open mind—both of which were soon put to the test when I learned the date of the ceremony. It was to be on the anniversary of Philippe's death. A solemn day on the Cousteau calendar. I could not understand how my father could have allowed this to happen. Or, maybe he had failed to make the connection, which in some ways would have been worse. And yet, here it was.

It was astonishing to me—but then, Dad was always full of surprises.

But for his second marriage? To go through these motions again when the bride was approaching her 50s, and the groom was on his way to 80? With their two children in attendance? Children who were young enough to be his grandchildren? It was a charade inside a puzzle—but this, too, was my father.

The charade lingered in my mind for some time. It did not trouble me so much as it perplexed me. I worked to understand it. I could only think that my father had lost his guiding star with the death of my mother, a loss that could surely shipwreck any navigator. La Bergère had been his compass, and she was no longer here to guide him; it was no wonder that there appeared a sudden lack of direction beneath his decisions or that his actions seemed to belie the man I knew him to be.

Yes, this was my father—a captain of men who could sometimes take a wrong turn.

And so it is with remorse and uncertainty that I look back on the final years of my father's life, as the "wrong turn" of his second marriage drove him further and further from the times we had shared in and around the world's oceans—and, further and further from me. I do not wish to revisit the many aches and pains of our relationship during this period except to give them mention. There was a roiling controversy surrounding my attempt to open an environment-friendly diving resort in Fiji, under the Cousteau name—a name that was of course mine as well as my father's.

There were the internecine struggles at the Cousteau Society, which fell more and more under the influence of the second Mrs. Cousteau, resulting in my ultimate departure from an organization I had helped to launch with my father and brother—an organization that for a time had seemed to encapsulate the ideals of my family. And, a growing wedge seemed to surface between me and my father, as his wife began to hamper our attempts to see each other—and, as our opportunities for professional collaboration appeared to diminish. For so many years, JYC and I had been partners, in work and in life, and now without realizing it, we found ourselves more like the disillusioned associates at the dissolving end of a long business relationship. We were connected by our history and our shared ongoing interests but pushed apart by circumstances and influences we could not readily understand.

I suppose the silver lining to Dad's final years was that he remained fairly active until his death. His days were busy and purposeful. He tried to keep up with his friends, with his diving, and with his advocacy on behalf of the planet. From time to time, our paths would cross; we would alight on the same project, at the same time, and come together briefly in the spirit of reconciliation.

The very last time I visited with him when he was healthy and alert and much like his old, gregarious self was at a diving convention in Orlando, Florida. I hadn't been certain that I would see him there, but I was hopeful we would find some time for each other. Happily, JYC was on his own for this trip, as he gave what would turn out to be his last official speech. I listened with great admiration as he spoke so passionately about the sea and its inhabitants and our responsibilities as caring, thinking humans to look after these precious resources. He was always a gifted and fearless speaker. In French or English, it didn't much matter. He could mesmerize an audience with his gifts as a storyteller—which, after all, was what he had set out to become in the first place.

Afterward, I joined him at his table, to reminiscence and reflect. We talked as father and son, as mentor and protégé, as former partners. Mostly, though, we could only talk around our various difficulties, and at one point, my father made a veiled reference to the lawsuit that had nearly derailed us. He lifted his glass of wine and said, "You know, Jean-Michel, when you have a problem, you don't need to call in your lawyers. You can get together over a bottle of wine and solve everything."

This had been his approach to life all along, and yet we never managed to share a bottle to discuss our disagreements over the Fiji resort, or the direction of the Cousteau Society, or his hurtful treatment of my mother. We did no such thing, and I had to believe that his remarriage was in part to blame. Did that mean my father was not responsible for the recent messes that lay in his wake? Not at all. Did that mean I could forgive him his trespasses? No, I didn't think so. I was not inclined to give my father a free pass, and neither would he have thought to ask for one. This was simply his sidelong way of acknowledging that he had made some mistakes in his life—just as I had surely made some mistakes of my own.

And so we drank.

It was the last time we were able to reach back over the years for what we had once taken for granted. When I saw him next, a couple of months later, my father was in decline. I had learned in a round-about way that he was in the hospital. If I called to talk to him, to see how he was doing, I was usually told he was resting. He could not be disturbed. Still, I kept calling—although I could not have spoken to him in any kind of traditional sense. I did not know this at the time, but he had slipped mightily over these past months. He couldn't talk. He could hardly breathe. At one point, he needed a tracheotomy, and this strong, proud man of the sea grew weaker and weaker.

My cousin Jean-Pierre, the same Jean-Pierre who was so much a part of my growing up as a young boy in the Alps and in and around Toulon, was now a well-known cardiologist in Paris, and my father was his patient. It was Jean-Pierre who ultimately alerted me to my father's hospitalization and kept me apprised of his condition, even as it worsened. One day, Jean-Pierre noticed an item in the newspaper, reporting that JYC was resting comfortably at home when in fact he was in the hospital in Jean-Pierre's care, so he called me to clarify.

I flew to Paris in the company of my good friend Tim Trebon, who did not believe I should make such a trip alone. Together, we raced to the hospital, but I could not get past security. Strict instructions had been given that I was not to be allowed in to visit. Me! The eldest son of Cousteau! It was a kind of outrage, but I did not have it in me to feel outraged. Instead, Tim and I retreated to a restaurant by the train tracks near Porte d'Auteuil and got drunk. I did not know what else to do.

A few weeks later, my cousin Jean-Pierre mercifully made arrangements for me to be allowed into Dad's room at a different hospital, where he had been moved. I was not prepared for what I found on the other side of the door. There he was, the great Captain Cousteau, looking frail and helpless. He was lying comfortably on his back, but his breathing appeared labored. A great many tubes and lines ran in a great many directions, from his withered body to various machines at his bedside.

And yet his eyes were bright and smiling. He was happy to see me, I could tell. He had something to say to me but no good way to say it, so he indicated with his eyes for me to take his hand. And so, I did. He held my hand and gave me a gentle squeeze, and then another. I squeezed back. But then, so did he, and as he did so, I realized this was the only way left for the two of us to communicate. Through Morse code! The way we used to do a half century ago—with dots and dashes!

He could hear me just fine, though, so I pumped him with questions. I said, "How are you feeling, Dad?"

And then, painstakingly, he responded in code, with short and long squeezes, spelling out his answer. "Tired," he managed to say. And then, "Sad."

"I am sad, too," I said, "but let us not be sad."

It took a while, but he managed to squeeze out another few comments:

"I don't think, I'm going to make it," he squeezed out.

"Don't say that," I said. "Don't believe that."

"It's been a grand adventure," he squeezed out.

"Yes, it has," I said.

"It is a shame we haven't been able to spend more time together," he squeezed out.

At this, I could only shed a single tear.

It was a very emotional, bittersweet moment, and I hated to have it end, but JYC grew tired soon enough.

The next time I saw my father was also the last. He was back at home. I got a phone call in London, from Dad's new father-in-law, telling me that JYC was near death. "If you want to see him, you must hurry," he said. So I hurried. When I arrived at his apartment a few hours later, I was met at the door by my father's attorney. My first thought was, "Oh, no, I am too late." But I was not too late. The attorney had simply been dispatched by my father's wife to monitor my visit. It must have been a difficult, awkward assignment, but he was kind enough about it.

He escorted me to a room where my father lay on a hospital bed, completely unconscious, surrounded once again by all manner of tubes and machines. I had never been to that room before, so I allowed myself a brief look. There were family photographs all around, all over the walls. So many photos—but they were pictures of Dad and his new wife and his young children, Diane and Pierre-Yves. There were no pictures of La Bergère, no pictures of the *Calypso,* no pictures of Philippe, no pictures of me. It was as if we had been blotted from my father's life like a small nuisance.

After a moment, the lawyer excused himself from the room. I guess he determined that I was not about to do anything objectionable. I appreciated the privacy, but I could not think how to fill it. My father was unconscious. I could hear his labored breathing. All around were photographs and memories that had nothing to do with me or my family or the life and adventure we had all shared.

I could not stay long. I did not wish to be in the apartment upon the return of his wife. I chose to remember my father as I had known him. I said my final good-byes, and made for the door.

LEGACY

The more I look back on my father's life and work, the more I realize what a visionary he was, even though he would have never used the term to describe himself. In the beginning, he was not much interested in the future. Rather, he was concerned with the present, enjoying himself to the fullest and experiencing the riches and wonders of the deep in the most consuming, most immediate ways he could imagine—and, if such a thing wasn't possible, he would find a way to make it so.

I suppose it can be said that he lived for the moment, at a time when such a sentiment did not even have a name. However, all those "moments" added up to something quite—well, grand. Sustaining. Everlasting, even. You see, Dad was a pioneer who broke barriers with his inventions such as the Aqua-Lung and the underwater camera, but as those barriers fell away, he could not help but see the consequences. Over time, he began to understand the risks of nuclear technology and waste to our precious underwater habitats. Over time, he could project the devastating results of overfishing and climate change. Over time, he started speaking out about our runaway population growth and the resulting strain on our natural resources. And, over time, people started to listen.

Without really meaning to and without realizing it, JYC became a kind of global ambassador of the sea—an ombudsman for the planet. The Cousteau "brand," if we can even label it as such, came to stand for the preservation and appreciation of the world's oceans, while Dad himself emerged as a kind of spiritual guide for the environment. In the years since his death, I have heard him described as an eco-warrior, a prescient demigod, a fierce advocate—terms he would have disavowed because he was none of these things in his mind, even though he was all these things and more in the eyes of the world. Still, this enduring public persona was and remains a happy outgrowth of my father's earliest free dives in the Mediterranean, and I believe it's important to recognize this connection. He did not set out to change the world, merely to experience it—fully and intimately and without constraint. All of these other benefits and associations simply attached themselves to my father's pursuits in such a way that they became intertwined, and soon it was impossible to separate the adventurist from the naturalist, the man from the mission.

"People protect what they love," he used to say—and just as his love of the sea was sure and certain, so, too, was the responsibility he felt toward it.

I am reminded here of an unexpected friendship that developed among JYC and his *Calypso* crew with a giant river otter we rescued from an abandoned zoo near the river. We had not expected to take this creature on board, but it was soon a cherished pet. This alone was surprising, because she was smelly and wet and unable to sit still. We called her Cacha, and she was constantly jumping into our bunks or onto the table, as much a pest as a beloved member of the family. Dad, ever the showman and ever mindful of the documentary cameras, had likely brought this river otter into our midst for comic or dramatic effect. But as he prepared to return Cacha to her natural environment, in what would emerge as a signature scene in a documentary on the Amazon produced for Ted Turner's TBS network, he took painful notice of the devastation all around. It was 1982, and the rain forests were already disappearing. Trees were being cut down at a staggering rate. Human enterprise and greed were threatening the lush ecosystem of the region, which my father had come to regard as the heartbeat of our planet. After all, the waters of the Amazon circulate through all the world's oceans, and we had only to look at the ruin and wreckage of the region through the eyes of our pet river otter to appreciate the dangers that lay in wait.

Yes, we become fiercely protective of what we love, and here we were moved to protect our pesky friend. The moment marked a kind of turning point for me, because I never again looked at my surroundings from a place of innocence. There would still be joy and wonder and adventure, but from this moment on, I was on a kind of campaign. JYC, too.

This was not the first example of my father's advocacy. As early as the 1950s, I would hear him grouse to his fellow mousquemers that the Mediterranean was becoming a garbage can. It was a sentiment he would echo for the rest of his life, and it was around this time that he organized a peaceful protest against the dumping that was becoming prevalent throughout Europe. While the men were off at work, he enlisted a village of women and children to be stationed up and down the Rhône Valley through the middle of France and all the way down to the Mediterranean in Marseille, and he positioned them along the railroad tracks, to prevent the French government from transporting a trainload of radioactive waste and dumping it into the sea. The protest barely made the news, but it was nonetheless effective. With all of those women and children on the tracks, the train could not get through, and the waste was rerouted and stored in a facility that was unofficially known in bureaucratic circles as the "Cousteau hangar"—not to honor my father, of course, but to blame him for derailing the original plans for disposal.

It was the first of many poetic gestures my father made on behalf of the planet, and he added another verse with each protest or shout of warning. In this way, the Cousteau Society emerged as one of the world's first and foremost environmental organizations, and we made it our mission to communicate our concerns to a global audience. Dad actually drafted his famous Bill of Rights of Future Generations as a clarion call to embody the principles of sustainability and responsible resource management. "Every person has the right to inherit an uncontaminated planet on which all forms of life may flourish," he wrote, and we eventually collected nine million signatures in support of this simple ideal. By 1997, the year of his death, the General Conference of the United Nations Educational, Scientific and Cultural Organization (UNESCO) formally adopted a reworked text of Dad's original draft, which was far more than a fitting tribute to the life and work of Jacques-Yves Cousteau; it was also (and primarily) a strong statement of purpose for all of us, for all time.

Today, it's not an issue of protecting the fish and the birds and the butterflies and the trees. It's about our very survival as a species and the survival of our ocean planet. "How can you protect what you don't understand?" my father used to ask. Well, when he was a young man, we did not understand such matters, but we do now. We understand that as we add another 100 million people to this planet each year, there will be an additional drain on our resources. We understand the real and pressing need to control our population growth. We understand the devastation taking place on our ocean floors. We understand that we are taking more from nature than nature can produce. We understand that we need to consider the preservation of the environment as we would any other business venture. It might not be a romantic view, but it's certainly pragmatic. If you look upon our environment as our species' greatest, most significant asset, then we must manage it like capital. We must realize that it's possible to live off the interest of that capital indefinitely as long as we take care of it. But if we start to erode that capital—well, then we'll careen toward bankruptcy. It's a simple equation, basic to any ongoing concern, as any economist could attest.

When it comes to our oceans, we're most assuredly taking away more than we're putting in and therefore depleting our assets. More and more, we use the sea as a sewage facility, dumping waste and chemicals that affect plant and marine life in such a way that the impact can now be felt all the way up the food chain. Even dominant species like killer whales are becoming sick, and in certain parts of the world, such as the Pacific Northwest, baby whales are dying at alarming rates. Soon, and in turn, we will become sick as well, because we will continue to put the contaminated fish on which those mammals feed onto our plates as well. Forget that the oceans are no longer so pretty to look at; they're dangerously approaching the point where they'll no longer be able to take care of us.

As he took up his advocacy for the oceans later in life, Dad found a variety of ways to make his point. The most compelling of these were the pictures he brought back from his various expeditions, but he also had a way with words. He used to talk about the planet in terms of an airplane—calling, perhaps, upon his own youthful leanings toward flight. He said, "If you remove one species, one environment, one habitat, it's like removing a rivet from an airplane. Remove just a few and the plane will still fly. Remove a few more, and you still might be okay. But at some point, you'll remove one rivet too many and the plane will come crashing down."

It is an apt metaphor, wouldn't you agree? And here we are, still okay, still flying, in desperate need of tightening our few remaining rivets and setting things right. It's not too late for us, I don't believe. We've had our wake-up call, and now we must do just that—wake up! The pressure is on. The moment for action is upon us. In the past, it took millions of years for the disappearance of a species to occur, whereas the rapid climate changes of today have made it so it can now happen within decades. We're responsible for accelerating that timetable, but we are also responsible for returning to our baseline norms. We must adapt to survive, and I am confident we will do so. But how much more will we have to lose, before we answer our wake-up call? How many people will have to move away from the coastline because their coastline is no more? How many more will starve?

We're running out of excuses. And time. It used to be that there were people and places on the planet that were totally disconnected from people and places elsewhere. In the remotest reaches of India or Africa, for example, it was possible to find tribes of people living much as they had for centuries, with no contact to the outside world. Today, rich or poor, sated or starving, we are all connected. Owing primarily to the advances in our information and communication technologies, we are one people at long last. And yet we are often at cross-purposes. We carry passports, which in some ways only serve to divide us. After all, we are one people, sharing one planet. Whales and dolphins do not have passports. From a philosophical point of view, the concept of passports for humans is fairly absurd—but, alas, we do not live in a philosophical world. We live in the real world, but let us not take ourselves too seriously. Let us remember that we showed up on this planet about three million years ago. Whales and dolphins were here 50 to 60 million years ago. Sharks go back more than 200 million years. We are but infants on this planet, really, and yet because of our formidable tools, because of our hands and our brain—and, most important, our imagination—we are in a position to determine our future.

If the planet goes to dust in a billion years, as scientists say it will and if a supreme being is asked to write the definitive history of Earth, will we humans even rate a mention? I'm not so sure. We're just a speck, a footnote, but at the same time, we have a tremendous capacity to effect change. We have a choice. We can become one with nature and embrace the challenges that face our planet, our oceans, our species. Or, we can continue along our present course to an uncertain future.

We've had it pretty good, but we've had it at our children's expense. We've lived way beyond our means as a species, and now we're headed toward bankruptcy. But we can change. We must. We will. I believe this deeply, fervently. If I didn't, I'd be out there, trying to catch the last fish. My father instilled the urgency of this belief in me. In fact, his personal evolution can be seen as a model for where we stand now—from that early protest in the Rhône Valley, to the first global environmental rally, the United Nations Conference on Environment and Development in Rio de Janeiro in 1992 to his ongoing participation at meetings and conferences all over the world and reaching all the way to the United Nations Climate Change Conference in Copenhagen in 2009, where he was surely there in spirit. These forums confirm that, as a species, we're not very good at anticipating a crisis, even as we appear boundlessly and tirelessly motivated by crisis. We humans are a reactive bunch. Now, we are surrounded by crisis, but the entire world is engaged. It's no longer just a group of bureaucrats or think-tank environmentalists or politicians. Now there are real, workable solutions being developed and sustained by real, workable initiatives—and there are more on the way.

My father was not perfect. In this, he was like all of us, which of course made him more than just a symbol of the sea. He had his flaws, and we had our disagreements. But his flaws and frailties make him no less of a symbol, and it is the symbol that endures. It is a memory I carry with me to this day: my father, my friend, my captain, my partner, my boss standing on the deck of our wind ship, *Alcyone,* looking ahead to what he might leave behind. He was an impatient man, who could also be stubborn and passionate and generous—all to a fault. And here he stood on the deck in the twilight of his life and presented me with an honor and a challenge, which belonged not only to me but also to each of us determined to be a part of a meaningful future.

He said, "It is you, Jean-Michel, who will carry the flame of my faith."

Yes, it is on me, just as it is on each of us, to carry on the work of this brilliant, difficult, complicated, charismatic man—to accept the challenges of today as the opportunities of tomorrow and to acknowledge the incredible privilege we all share on Planet Ocean.

Let us remember, when you protect the ocean, you protect yourself. My Dad taught me that.

TIME LINE

1910, June 11: Jacques-Yves Cousteau is born in Saint-André-de-Cubzac

1930: Enters Ecole Navale; graduates as gunnery officer

1936: Begins diving experiments with Philippe Tailliez (also a naval officer on the naval vessel *Condorcet)* using Fernez glasses

1937: Marries Simone Melchior

1938: Is sent on missions to Shanghai and Japan on the *Condorcet* as part of the information service in the French Navy

1938: Son Jean-Michel is born.

1939: Is sent on mission to the USSR

1940: Son Philippe is born.

1943: With Emile Gagnan, invents the "aqualung," which Cousteau, Tailliez, and Frédéric Dumas test

1944: With Tailliez and Dumas, the team of divers, nicknamed the mousquemers, is tasked with removing two torpedoes from a scuttled German submarine off the peninsula of Saint-Mandrier near Toulon and with mine recovery and detonation. Cousteau begins underwater filming.

1945: Under Tailliez's command, the Groupe d'Etudes et de Recherches Sous-Marines (GERS) is formed. The French Navy provides the ship *L'Eli Monnier,* Cousteau's first command, with Tailliez continuing overall charge of the group. The group invents more lifesaving equipment: the Alinat lifesaving belt and the Dumas buoy.

1946: Explores the Fontaine-de-Vaucluse, a freshwater grotto, with the group. They begin using "breathing cylinders" filled to 3,000 psi. Cousteau and Dumas nearly lose their lives in the testing.

1948: Conducts archaeological exploration of the *Mahdia* galley off the coast of Tunisia with the group

1949: Is involved in the rescue of Professor Jacques Piccard's bathyscape off the coast of Dakar. Cousteau leaves the group.

1950: Equips the *Calypso* and founds the Campagnes océanographiques françaises.

1951–1952: Begins expeditions in the Red Sea, which result in the publication of *The Silent World* in 1953

1952: Explores an ancient Roman ship at Grand-Congloué, near Marseille

1955: Embarks on a 13,800-mile journey, which is recorded for a film version of *The Silent World*. The film receives the Palme d'Or in Cannes in 1956.

1957: Retires from the French Navy and becomes director of the Oceanographic Museum in Monaco

1960: Organizes a protest against dumping nuclear waste in the Mediterranean

1961: Conducts Conshelf I experiments for underwater living, followed by Conshelf II experiments, which result in the film *World Without Sun,* shown in 1964 in the United States. President John F. Kennedy presents Cousteau with the National Geographic's Special Gold Medal for his invention of the regulator.

1964: Son Jean-Michel graduates with a degree in architecture from the Paris School of Architecture

1965: Son Philippe graduates from l'Ecole Technique de Photographie in Paris, the French government's school of motion pictures

1966: Begins work on the first hour-long special of *The Undersea World of Jacques Cousteau* on U.S. television

1968: Launches the premiere of *The Undersea World of Jacques Cousteau* on ABC, starring his sons Jean-Michel and Philippe

1973: Founds the Cousteau Society for the Protection of Ocean Life with his two sons and Frederick Hyman

1977: Premieres *Cousteau Odyssey* series on PBS, expressing concern about the environment. Together with Peter Scott, Cousteau receives the UN International Environment Prize.

1979: Philippe Cousteau dies in a seaplane crash. Jean-Michel takes his place alongside his father; they collaborate for the next 14 years.

1980: Begins work on the turbosail to launch the ship *Alcyone* in 1985. Cousteau also experiments with *Argyronète* (Sea-Spider), a device to analyze the biochemistry of the ocean's surface.

1984: Premieres the *Cousteau Amazon* series on Turner Broadcasting System

1985: Receives the Grand-croix de l'Ordre National du Mérite from the French government as well as the U.S. Presidential Medal of Freedom awarded by President Ronald Reagan

1988: Becomes a member of the French Academy of Letters. That same year he is honored by the National Geographic Society with its Centennial Award for special contributions to mankind throughout the years.

1990: Simone Cousteau dies of cancer.

1991: Marries Francine Triplet with whom he already has two children

1992: Is invited to Rio de Janeiro, Brazil, for the United Nations International Conference on Environment and Development

1997, June 25: Jacques-Yves Cousteau dies in Paris.

Acknowledgments

M y parents have passed away. My brother has passed away. Today, *je fais le point* (I am taking stock). I particularly wish to express my gratitude for the talent of Dan Paisner for allowing me to reestablish the truth and to release what I am carrying in my heart.

A special thanks to Jean-Daniel Belfond, Barbara Brownell-Grogan, Karin Kinney, and John Silbersack.

I also wish to express my deep and emotional gratitude to all the crewmembers of the *Calypso* and *Alycyone*. Their courage, dedication and tenacity are a tribute to my father's work. They unquestionably gave their support to La Bergère, without whom my father would have never been able to fulfill his mission.

From the bottom of my heart, I wish to thank my children, Fabien Cousteau and Celine Cousteau; my cousin, Jean-Pierre Cousteau; Captain Jean Alinat; James Andrews; Michel Deloire; Frédéric Dumas; Jacques Ertaud; Albert Falco; André Laban; Titi Léandri; Nancy Marr; Dr. Denis Martin-Laval; Dr. Richard Murphy; Celso Oliveira; Marie-Claude Oren; Elise Otzenberger; Christian Petron; Anne-Marie Roth; Jean-Charles and Monique Roux; Don Santee; Dominique Serafini; Sandra Squires; Pam Stacey; Captain Philippe Tailliez; Alain Traonouïl; Tim Trebon; Charles Vinick; Carrie Vonderhaar; and everyone at Ocean Futures Society (www.oceanfutures.org) for their precious contributions.

My thanks as well to all the others who may identify themselves with the work of Jacques-Yves Cousteau.